Digital Literacy

Susan Wiesinger & Ralph Beliveau

Digital Literacy

A Primer on Media, Identity, and the Evolution of Technology, Second Edition

PETER LANG

Lausanne • Berlin • Bruxelles • Chennai • New York • Oxford

Library of Congress Cataloging-in-Publication Data

Names: Wiesinger, Susan, author. | Beliveau, Ralph, author.
Title: Digital literacy : a primer on media, identity, and the evolution of
technology / Susan Wiesinger, Ralph Beliveau.
Description: Second edition. | New York : Peter Lang, [2023] | Includes
bibliographical references and index.
Identifiers: LCCN 2022058049 (print) | LCCN 2022058050 (ebook) |
ISBN 9781636671000 (paperback) | ISBN 9781636671017 (ebook) |
ISBN 9781636671024 (epub)
Subjects: LCSH: Information technology—Social aspects. |
Information society. | Digital media—Social aspects.
Classification: LCC HM851 .W545 2023 (print) | LCC HM851 (ebook) |
DDC 303.48/33—dc23/eng/20230113
LC record available at https://lccn.loc.gov/2022058049
LC ebook record available at https://lccn.loc.gov/2022058050
DOI 10.3726/ b20561

Bibliographic information published by the **Deutsche Nationalbibliothek**.
The German National Library lists this publication in the German
National Bibliography; detailed bibliographic data is available
on the Internet at http://dnb.d-nb.de.

Cover design by Peter Lang Group AG

ISBN 9781636671000 (paperback)
ISBN 9781636671017 (ebook pdf)
ISBN 9781636671024 (epub)
DOI 10.3726/b20561

*Financial support was provided from the Office of the Vice President for Research and Partnerships and
the Office of the Provost, University of Oklahoma.*

© 2023 Peter Lang Group AG, Lausanne
Published by Peter Lang Publishing Inc., New York, USA
info@peterlang.com - www.peterlang.com

Table of Contents

APPLIED SKILLS APPENDICES

Acknowledgements

Susan Wiesinger extends her limitless appreciation for John Wiesinger, an exceptional first-line editor and keepin' it real content evaluator. As Kenneth Burke would say, *without whom not*.
Additional thanks to:

- The tech journalists and academic researchers who keep close watch on tech trends and cultural impact. Your work is extraordinarily valuable and we're proud to have aggregated some of it in this book.

- The students of JOUR 255, "Digital Literacy & Media Technologies," at California State University, Chico, who helped refine the work and offered feedback on final chapters. (Beyond that, they are an endless source of inspiration.)

- Regan Penning and Sofia Thomas, who proofread the manuscript as students in the above-noted class, taking the "typo bounty" seriously enough that I came to think of them as my editorial assistants.

- The anonymous reviewers whose time and energy reviewing earlier versions of the manuscript greatly strengthened the final book.

- The University of Oklahoma, which provided grant funding for licensing of chapter comics.

- Vin Crosbie, who provided extraordinary feedback of an early draft of Chapter 2, inspiring the author to do more.

- Josh Boyd, who nurtured my research and writing as a student at Purdue University and made me think I could do *all* the things. More than two decades later, I feel that I have, which I attribute to / blame on his influence.

Ralph Beliveau says this work was dependent on Abby and Martha, who continue to shame him about his media experience … or lack thereof.

Additional thanks to:

–   Laura, the best teacher anyone could ever have.

–   Marilyn Beliveau, for making learning a great thing.

–   Mike Dolesh for giving me troubling things to read when I was impressionable, which made me a better person. He was the best teacher I ever had.

Any mistakes are the sole responsibility of the authors.

What's New in the Second Edition

Digital literacy is a question that keeps changing, and much has changed since the first edition of this book.

In drafting the second edition, we've reviewed, updated, and reorganized, as well as added a significant amount of new content around the concepts of media ownership, consequences of Big Tech dominance, impact of social media, individual engagement, physical impacts of ubiquitous tech use, global effects of boundary-crossing information, environmental costs of tech, media literacy, and social costs of disinformation.

For example, the spread of COVID-19 and subsequent slow burn of a pandemic marked a significant disruption to our collective worldview and sparked deeper analysis of the consequences of our reliance on mediated communication technologies. During the infamous lockdown of spring 2020, our in-person connections with friends, family, coworkers, teachers, and classmates were abruptly stopped. Of course, we all know the plot of this particular story: video teleconferencing and streaming video to the rescue.

While we'll weave discussion of digital literacy and the pandemic into various chapters ahead, it's worth noting here that we learned at least a couple of things about ourselves:

1. Mediated contact, whether face-to-face via Zoom or scrolling through social media feeds is a not a satisfactory replacement for human interaction for everyone; and,
2. We crave at least the semblance of common culture.

The former has emerged as a collective increase in anxiety and lost learning opportunities that will affect our communities for years to come, while the latter provided lighter moments of connection along the way—from cursing Zoombombers to collectively binge watching "Tiger King" and multiple seasons of "Schitt's Creek" from within our lockdown bubbles. The conversations that resulted on social media made us feel like we were sharing *something* in the midst of significant uncertainty.

We're also hoping the second edition piques interest in the growing body of research around the effects and impact of the digital technologies we use every day. We've embraced the second edition as an aggregation of some of the best tech journalism and research available—critical work by knowledgeable folks that is worthy of your consideration.

CHAPTER 1

Introduction: The Book Starts *Here*

Figure 1.1. *What does it mean to live in an information landscape where filters—aka journalists—are no longer the only people to interpret, document, and draw the public's attention to the issues and events that continually shape and change our world? Pearls Before Swine © 2005 Stephan Pastis. Reprinted by permission of Andrews McMeel Syndication. All rights reserved.*

OK, honesty time: How much do you read? We mean really *read*—as in sit down with a single means of information delivery (aka a book, magazine, or newspaper). And how often are you criticized or feel guilty for not reading enough?

A hallmark of digital living is that you likely are *consuming* more information than any previous generation, but you're not necessarily *reading* it. We're all skimming, skipping, and surfing as we navigate an endless ocean of content via a veritable flotilla of media and digital device choices.

The flood of information always has been there, but prior to the Internet it took owning a printing press or having access to the airwaves to transmit information to the masses. In an information landscape dominated by print and broadcasting, only certain people could participate. And those people—mostly journalists—were held to particular newsgathering practices and values that effectively excluded the public.

Not surprisingly, when the internet opened participation to the rest of us, many took the opportunity to put their unvarnished observations, announcements, meals, and minutiae directly online, cutting out the traditional mediator. And so, the floodgates fell and the information tsunami came roaring through.

Media Literacy, Digital Literacy ... What's the Diff?

It's possible your grasp on the need for media literacy is somewhat weak, yet you're now being asked to grapple with digital literacy. That means there's a bit of catching up to do.

Media literacy is the ability to think critically about the endless flood of messages sent by media organizations (e.g., films, shows, news reports) to a general, undifferentiated public, as well as those we receive via our highly individualized social media feeds. It's all about how we interpret and respond to those messages—consciously or unconsciously.

Traditional media literacy assumes a *passive* population that consumes a significant volume of messages, from which a **common culture** is built. Common culture emerges when there are touchstones—shared, collective experiences—that bind us together as a society.

Think of it like **water-cooler culture:** People taking a break, standing around the office water cooler, and talking to coworkers about a television show that everyone watched last night. If you didn't watch, you're left out. And next week you watch so you can share in the water-cooler conversation.

Digital literacy is different, largely because *technology* has changed how we communicate, what we consume, across what platforms we consume it, and via what devices. Digital literacy also is concerned with our interpretation of and response to messages, but is more focused on the cultural impact of the technologies delivering the information.

Digital information typically involves both passive consumption and active production of highly personalized messages. It holds the promise of being interactive, where producers become audiences for users, and users produce content for each other. But from a historical perspective, the result is both fragmentation and the erosion of common culture.

Think of it as **spoiler culture:** People still take work breaks, but they don't leave their desks (or sometimes even their homes). Instead, they turn to digital tools that enable the near-instant transmission of cultural symbols from person to person via internet-connected digital devices.

Because we are all consuming information across different channels, we don't share reference points that are anchored in time and space. For example, you may choose when to watch your favorite television show and via what medium, rather than being held to a network schedule and required to watch on a television set. "Gilligan's Island" from 1964 on Hulu? Check. (Just don't tell me how it ends.)

The difference between water-cooler culture and spoiler culture—media literacy versus digital literacy—is the method of transmission.

Instead of person-to-person and media-to-masses, spoiler culture is a digitally mediated world of individuals. Even if every person were on the same social media platform, it is unlikely that common culture would emerge. That's because it's not the platform that matters; it's the information feed. And every digital feed is unique. Each is customized to reflect the user's personal interests, values, history, and increasingly, brand.

In the past, television was a social medium that encouraged people to watch shows together. Even if you weren't in the same household, you could share the experience of having watched the same show (aka common culture).

Digital literacy means understanding "television" as both a social and a solitary thing to do. Sometimes we gather around the TV. Sometimes we watch on our digital devices alone—or perhaps even chat with other people on similar devices in different places at the same time.

It's increasingly rare for individuals to watch what everyone else is watching or even consume information across one digital device at a time. In fact, you may be physically in a room watching TV with others, while skimming content on a laptop or tablet and interacting with others on a smartphone.

For that matter, television no longer exists as a stand-alone cultural norm for a decent chunk of the U.S. population, particularly those who have given up, or never signed up for, cable TV. A great deal of what might be termed TV content now is consumed on digital devices, including TVs, accessed via video streaming services.

If you've ever watched a made-for-TV series via a streaming service, you'll note that each episode is the same length and there are logical points where advertising would have been placed. Those shows were created for a world bound by time.

If you watch a made-for-streaming series or user-generated video on YouTube and choose not to pay for an ad-free service, you'll note that the episodes can vary wildly in length and ads may be dropped in at the most inopportune times. They're not pauses built into the plot, they're payment for watching. Those shows were created for a world bound by individual preference.

Existing and emergent technologies have put more of *everything* at our fingertips and in our brains than ever before. We have not discarded previous media as new have arisen; instead, we have accumulated—print, film, radio, television, cable television, internet, World Wide Web, social media, apps, and so on.

A digital information environment is one in which the *medium*—the device through which we access information—has a larger social impact than the *message*—the content we receive through a particular medium.

As a result, we need to critically consider what it means to be a citizen in an information landscape that not only has been individualized but also is constantly in flux.

We also need to understand that consumables—from user-generated content to social media feeds to streaming video—have increased exponentially. And much of that content is being consumed online and alone.

One concern with information inundation is that people no longer will be interested in what is real or even relevant to their physical lives, disrupting our ability to reason and participate reasonably in a democratic society. Media theorist Neil Postman speculated that we would become increasingly immersed in "a peek-a-boo world, where now this event, now that, pops into view for a moment, then vanishes again."[1]

Postman wasn't opposed to entertainment, but he was conscious of increasing levels of distraction. He understood that people need the escape that entertainment offers, but he predicted that the rise

---

1   Postman (1985), p. 77.

of electronic media and increased focus on entertainment could have a two-fold effect on common culture and civic participation. In summary,

1. People wouldn't know what actually affects their lives; and,
2. They wouldn't know what they should do, even if they could figure it out.

If No One Is Gatekeeping, How Do We Know What's True?

In the world that print once ruled, the mass media tended to be assessed as purveyors of "capital-T-Truth," as in there is a Truth that can be discovered, verified, and agreed upon. This makes some sense, given the heavy use of newspapers as historical archives. The fact that print stories were finished and frozen in time granted them a degree of credibility not possible from any other medium.

Digital reality is more complicated, largely due to the volume of information being contributed from a multitude of voices. Because both journalists and the public are participating, this information flow often is viewed as "lowercase-t-truth," which implies that most observations are not *the* truth but simply one perspective from one point of view at one point in time. Or, as with Wikipedia, a *negotiated* perspective.

While this muddies source credibility on many levels, *we* don't necessarily want legacy media to help out with this problem by **gatekeeping** the flood of information.[2] *We* don't want the media to tell us what to think about, but want to discover the relevance to our lives ourselves. *We* want highly personalized content and to have the ability to share, challenge, and fact-check it in real time.

Paradoxically, content no longer is filtered exclusively by journalists but—by design and necessity—is filtered by complex computer algorithms that underpin browsers, search engines, and social media.

In short, our content is still subject to gatekeeping, but by the platform media companies that act as our conduit to the content—most notably the products of Alphabet (YouTube, Google) and Meta (Facebook, Instagram).

This type of dissemination gives corporations great power over individual information consumption and privacy, even as it allows people to think they are choosing what information they want to consume and from whom. It's also indicative of our existence in a **post-narrative world**, where overarching stories, an informed electorate, and the quest for Truth are being challenged by personal opinions, divisive ideologies, and individual truths.

As a result, much of what we consume reflects an algorithm-constructed impression of our own values and beliefs right back at us. We chat, message, comment, text, and like; we post, pin, subscribe (as long as it's free), and share—effectively witnessing everything from snacks to global politics through the screens of digital devices.

And this is an environment that allows individual users to share their knowledge—and their biases—to a much wider audience than ever before.

One need only think of Wikipedia to understand the ways in which the web allows users to collaborate and build deep reservoirs of knowledge about an endless range of topics. But as any high school teacher will tell you, Wikipedia is a starting point, not a definitive source of information—and

---

2   The gatekeeping role of the media was identified in a now-classic 1950 study by David Manning White entitled, " 'The Gate Keeper': A Case Study in the Selection of News." The basic idea is that the role of the journalist is to filter the news of the day. This idea is inextricably tied to the assumption made by Walter Lippmann in 1922 (and later with agenda-setting theory) that the media don't necessarily tell us what to think, but the filtering and channeling of news establishes what we think about.

you should not always believe what you skim. This is because people with a specific ideology can slant information by inundating the marketplace with their ideas or overwriting the ideas of others.

For example, prolific Wikipedia contributor Bryan Henderson, who goes by the name of "Giraffedata," took it upon himself to eradicate use of the term "comprised of" from user-generated content on the popular information site.[3] Over the course of eight years, Giraffedata removed the term more than 47,000 times in user postings.

Was this a grammatically sound move? Arguably, yes. But the fact that the "error" was made in more than 47,000 posts (and counting) speaks to the fact that the use of "comprised of" is common and acceptable.

Why did Giraffedata remove the term? Because he could. And, so can you. You don't like the color "yellow"? You wish broccoli didn't exist? Feel free to edit it out of your digital universe.

Because digital information doesn't *have* to flow through a gatekeeper, it's probable that it *hasn't*. And that opens up our sources of information, but complicates the quest for credible information.

Where Are We Going? And How Are We Going to Get There?

This is where digital literacy comes in. Awareness of the influence of the media, in general, and the profound impact of digitally distributed information, in particular, can help you navigate both current trends and future change.

In short, gaining a better understanding of your digitally filtered world gives you more control over it. To help you get there, this book approaches digital information literacy through three lenses:

> An **historical perspective** reviews snapshots of time and space to delineate how things were to lend context to how they are.
> A **cultural perspective** explores how values and ideals are constructed and conveyed within a given cultural context—how humans absorb and share the informal rules and norms that make up a society.
> A **critical perspective** illuminates how social changes—particularly rapid ones—can put large groups of people at social and/or economic disadvantage.

All three are necessary to help you understand the myriad ways in which our very identities are being altered by technology. Our own power to understand and control is grounded in how we understand power and control in general.

The focus of this book is on the relatively rapid evolution from common culture to digital culture, from communities of place to communities of space, from identity to multiple and conflicting identities. It's from a communication-centered perspective, one that considers the impact of both traditional mass media—aka journalism—and contemporary social media.

Before you settle in and start reading—yes, *reading*—we'd like to share a little of the philosophy behind this particular book.

It's written in a style that considers how digital media have trained you to read—and the short attention span that comes from exposure to an endless array of information at your fingertips.

The modifications include:

– A more conversational, approachable tone.

---

3 "Don't you dare use 'comprised of' on Wikipedia: one editor will take it out" by Geoff Nunberg, NPR Fresh Air.

– Short paragraphs, bold keywords, and bullet points—structured a bit like the information you consume day in and day out online and intended to be supplemented by information you can search for online.

– Citation that is on the page, rather than at the end of the book or chapter. We'll call that instant gratification.

– Sources that include more than books and academic articles. We provide full on-page citations for some of the best tech journalists out there, so you can follow their work.

– Footnotes that add to the conversation, rather than just acting as citation.[4]

This is not a classic text-heavy textbook.[5] It doesn't pretend to hold all the answers, but it asks you to think about the questions. It's more of a guidebook, really, aggregating the best tech of journalism and drawing wisdom from those who study and research technological change.

And we think you'll be OK with that.

Summary

This book explores media, communication, and technological change from historical, cultural, and critical perspectives. It is a decidedly journalistic approach, one that focuses on how the information technologies we adopt affect our identities as members of cultures and communities. The book presents a broad overview of complex issues, which you then are free to explore at great depth via the vast digital information resources at your fingertips.

Key concepts

Common culture
Digital literacy
Gatekeeping
Media literacy
Post-narrative world
Spoiler culture
Water-cooler culture

References

Nunberg, G. (2015, March 12). Don't you dare use 'comprised of' on Wikipedia: One editor will take it out. *NPR Fresh Air.* Retrieved from http://www.npr.org/2015/03/12/392568604/dont-you-dare-use-comprised-of-on-wikipedia-one-edi tor-will-take-it-out

Postman, N. (1985). *Amusing ourselves to death: Public discourse in the age of show business.* New York, NY: Penguin.

White, D. M. (1950). The "gate keeper": A case study in the selection of news. *Journalism Quarterly, 27,* 383–390.

---

4   So, hey, read the footnotes. Often you'll learn something new; occasionally you'll be rewarded with full-on snark.

5   We're not pandering to short attention spans, but we are acutely aware that mixing pureed cauliflower into macaroni and cheese is sometimes necessary to get kids to eat vegetables. If what you are consuming tastes good, goes down easy, and is good for you, we all win.

CHAPTER 2

The Evolution of Contemporary Media: Recalling a Collective Past, Sharing a Fragmented Present

Figure 2.1. *More communication options don't necessarily translate into better communication. Non Sequitur © 2019 Wiley Ink, Inc. Dist. By Andrews McMeel Syndication. Reprinted with permission. All rights reserved.*

Takeaway: The mass media system has had a significant role in creating and conveying shared goals and common culture. It worked well, even as communication technologies evolved and accumulated—then along came social media and digital devices.

Take this scenario: You open your clothes dryer to discover that you left a pack of gum in the pocket of a pair of jeans. While the entire load of jeans is inexplicably clean and gum-free, there's a webbing of gum baked onto your dryer drum. Grab a wet paper towel and start scrubbing. Now there are bits of paper towel stuck in the gum.

While you could try countless ideas, the best one is to head for The Google. Typing "removing gum from dryer" into a search engine string results in approximately 829,000 hits—most of which have been contributed by people exactly where you are now. Five minutes later the gum is gone (a dryer sheet does the trick), and you can add your own gum-removal success story to the cacophony of the web—"removing gum from dryer" hit number 829,001. And, if you want to fall down a rabbit hole, do the same search on TikTok, where you'll get basically the same advice, but set to music.

Legacy Is a Scavenger Hunt; Digital Is an Information Hunt[1]

The preceding example points to a significant difference in information discovery between legacy media—traditionally framed as mass media—and hyper-personalized digital media.

For example, consider the experience of flipping through the Yellow Pages[2] to find a service versus putting key words into a search engine to find the same thing. On one hand, you're exposed to lots of other services you may not have known existed, by simple virtue of having to flip through and skim pages to find what you're looking for in a *preset, socially prescribed format*. On the other, you get search returns that fit the parameters *you set*.

How do you look up a word online? You put the word into a search engine and the definition pops up, along with synonyms and antonyms. You don't need a dictionary or thesaurus in hand; it's all at your fingertips. You'll also get links to anywhere from a few to hundreds of millions of pages that include the word, depending on how common its usage is. Given some time, you could learn a great deal about *one* particular word by consuming these resources.

How would you use a paper dictionary? You'd have to look the word up and flip through the pages. In doing so, you might discover words that precisely convey little bits of human existence—words you didn't know existed.

Use of legacy content—from flipping pages to flipping channels—exposes you to a variety of information that you might not otherwise stumble across.

For example, the storage technology of paper means that even if you are only interested in getting to your favorite news section of a newspaper, you have to physically move through the others to get there. And by doing so you are exposed to a range of headlines and photos that are designed to grab your attention.

When it comes to digital engagement, you may only go to native apps on your phone or book-marked favorite sites on your laptop. And if you do a Google search on a topic, you may end up in the

---

1   Portions of this chapter were originally published in Chapter 3, "The Evolution of Media Technologies," of *Media Smackdown: Deconstructing the News and the Future of Journalism,* by Abe Aamidor, Jim Kuypers, and Susan Wiesinger.
2   Hahahaha, when's the last time you saw, much less picked up, a phonebook? They're still a thing, apparently, as one periodically lands on the front porch. Ask your grandparents.

middle of a site at exactly the article you were looking for. By doing so, you bypass the opportunity to learn new information, choosing instead the option of immediate, focused information.

The allure of instant gratification has resulted in legacy media being replaced by search engines as we hunt for information that fits our immediate needs. But the hunt for personal information—*your need right now*—and community-building, collective work that helps shape the future are not necessarily the same thing.

Legacy Versus Digital Media

Journalism traditionally has worked to document a society's history and influence its values. It was within the pages of newspapers that our democracy first emerged, via the pseudonymous writings of people like Benjamin Franklin. And journalists have largely continued with the founding fathers' charge of maintaining freedom and democracy by helping voters make informed choices, following the trail of how tax dollars are spent, and covering the dealings of local, state, and federal governments. A free press, quite simply, is the foundation of democracy.

Journalism traditionally was conveyed via stand-alone media technologies: print or broadcast. For nearly 200 years, print news sheets, newspapers, books, and magazines were the only form of mass media in the new world. Then along came electronic media that delivered film to theaters and radio and television broadcasts into American homes.

From about 1700 to the 1980s, news, information, and entertainment were bound to the technology that delivered them. Such **legacy media,** which typically include newspapers, magazines, books, documentaries, TV, and radio, existed long before cable television, the internet, personal computers, the World Wide Web, and smartphones. Legacy media tend to share the following characteristics:

– *Finite content.* There is only so much information that will fit in a space-limited newspaper or within a time-limited broadcast. A newspaper, magazine, or book can be read cover-to-cover, a TV show watched in its entirety, a movie watched beginning to end.

– *Fixity.* The content, once completed/published/distributed, does not change. A news sheet from the 1700s today offers exactly the same content as it did when it was first published. The fixity of film is part of the medium's charm: When you rewatch a movie, the ending doesn't change.

– Focus on a *general, mass audience,* assuming that all needs and desires for information are the same (though what people actually do with them can vary wildly). For example, when you buy a newspaper from a newsstand or tune into the 6 o'clock news, you are getting the same thing that everyone else is.

– *Anchored to the physical world,* with key goals of community building and/or entertainment.

– *Not interactive.* These media provided a one-way flow of information from the organization to an audience. Even at points where audiences were heard from, their role was limited and managed—more along the lines of providing input after the fact, rather than participating in a conversation.

– *Platform specific,* which means that content consumption is intended via a single technology (e.g., paper, television, radio, movie projector).

- *Linear and contextual,* in that information is intended to be consumed in a proscribed order. Television programs and series have a story arc that plays out across the beginning, middle, and end. Books are to be read from beginning to end. News stories are to be consumed from the headline to the lead, with details building as the reader reads.

- *Appointment oriented.* Television and radio have a schedule you have to follow. Newspapers—even the digital version—come out at a particular time of day. Movies are screened at particular, published times.

- Created to encourage *public engagement* as a source of fact-based, confirmed information.

Use of the term "legacy media" to describe contemporary media is somewhat misleading, as it carries the negative connotation that such organizations are caretakers for "old media," despite their significant contemporary play in the digital realm. We can learn more about how subsequent media work by considering how this "legacy" carries forward into the newer media. But there are fundamental differences between legacy and digital media that do make a critical distinction between the two.

Digital media have existed since the late 1960s but didn't achieve true mass adoption until advent of the World Wide Web, browsers to view it, search engines to index its content, protocols for secure shopping, and social media to move it all around. Digital content today is accessed via the internet, World Wide Web, and apps, and consumed via desktops, laptops, and mobile devices. Digital media tend to share the following characteristics:

- *Infinite content.* It's impossible to consume the entire World Wide Web, for example.

- *Interactive, ever-changing.* The only way to ensure that a website, video, news article, social media feed, and so on, will be exactly the same the next time you look at it is to take a screen shot (which creates an image of the content, an artifact from a point in time), or download it in the case of audio and video.

- *Focused on the individual,* who actively selects content that meets current needs and desires.

- *Anchored to digital devices* that open windows into the virtual world.

- *Platform agnostic,* which means that consumption occurs across multiple platforms (e.g., smartphone, tablet, laptop, desktop).

- *Nonlinear and free-standing,* in that information is intended to be accessed in any order the user chooses and expectedly out of context from any overarching narrative. Information is consumed in chunks that meet immediate needs—a 15-second video, just the biography section of a celebrity's IMDb page, and so on.

- *Drop-date oriented.* Content is available online as soon as it's uploaded. While you may have to wait for the next episode of a streaming show to drop, you can watch it at 1 minute after midnight if it makes you happy.

- Evolving to cater to *individual interests* as an endless source intermingled information, opinion, and entertainment.

The distinction of legacy and digital media presents a ***false dilemma.*** This means that there is a spectrum between legacy-only and digital-only, but we need to pretend they are completely separate things so we can see more clearly where they began and where they are today.

The Rapid Rise of Online

To figure out just how quickly legacy media have been changed by the rise of digital media, it's important to understand how the world of information and communication has changed. We'll tackle this by dividing the evolution of digital media into five waves.[3]

The First Wave: 1984–1993

It is amusing, if not helpful, to categorize the waves of digital media in terms of the capacity of various weaponry to blow holes in stuff.[4] The first wave thus can be categorized as a **bazooka approach**, in that the emphasis was on the mass audience.

Now, imagine that mass audience as a large, red barn, with the bazooka taking out a remarkably large chunk of the barn on the first shot.

Much as the bazooka is designed to blow big holes in things, first wave media technologies were taking aim at the audiences of early 1980s cable television and, sticking to the traditional mass-market model, were trying to reach pretty much *everyone* with the same message.

In the first wave, newspaper companies partnered with AT&T to send news via the phone lines to boxes hooked up to TV sets. The news, all text and presented in glorious black-and-white, would scroll down the screen.

The system had the advantage of taking control of the television, which prevented viewers from watching cable, but it had the disadvantages of being unappealing and slow.

The first wave also marked the introduction of chat rooms. In 1984, Apple's Mac had just come out and the personal computer was beginning to get a toehold in the market. The same newspaper-AT&T partnership came up with the text-based online service Viewtron, which allowed people to use their personal computers to interact with other people via a modem-to-modem computer network.[5] The system connected users to newspaper-hosted networks via the phone lines. These community-based networks attracted thousands of users at their peak.

The goal was for people to discuss the issues affecting their communities, things they *should* want to talk about, while also being exposed to text-based advertisements. What the media partnership learned, however, was that there was less interest in news feeds than there was in the media-hosted chat rooms. Discussions tended to be less about public affairs and more about personal ones, things people *wanted* to talk about rather than what civic-minded folks *should* talk about—in particular, sex, sports, and entertainment. The system was shut down.

By 1986, a handful of companies, including America Online (AOL), Prodigy, and CompuServe, had stepped in to fill the void, offering home internet access and basically saying, "Hey, if you want to come over and be anonymous and talk about sex, be our guest." These commercial networks—popularly known as Online Service Providers—attracted tens of millions of paying subscribers.

Upshot: Customers rule. These companies listened and offered end users exactly what they wanted.

---

3   This is an expansion on the three waves of online mapped out by *Online Journalism Review* editor Larry Pryor (2002), who was drawing on the work of web consultant Vin Crosbie and scholar François Bar.

4   In presenting this analogy to colleagues and college students in the European Union, we were met with the comment that Americans really do build violence into everything they do. Interesting observation, worth noting but beyond the scope of this book.

5   "Modem" is short for modulator-demodulator, which is where conversions from digital signals to analog signals happens. You can't really hear or see a digital signal; it needs to be converted to images and sounds.

As a kind of "bookmark" for this first wave, consider a 1992 work titled *Agrippa*, which was written by Science Fiction author William Gibson (who invented the term "cyberspace") and artist Dennis Ashbaugh.

You could get *Agrippa* as either a 3.5-inch floppy disk or an art book. But as you read the text from the disk, it encrypted itself,[6] disappearing, and the book was chemically treated to gradually fade when exposed to light.

The Second Wave: 1993-2001

The second wave of digital media can be categorized as a **shotgun approach,** in that the emphasis was on *mass demographics* in an effort to reach parts of the greater audience.

Keeping the same large, red barn in mind, think of the shotgun as being able to target various parts of the barn with varying degrees of accuracy.

The second wave was facilitated by the invention of the World Wide Web, but it still was a text-based system. The first web browser, aptly named "WorldWideWeb," was available to the public in 1991 and, by 1993, about 1.3 million computers were connected to the internet.[7]

But the web didn't really take off until developers launched Mosaic, a user-friendly browser that could translate World Wide Web code into images that appeared on websites alongside text, rather than downloading as separate files, and clickable links. This graphical hypertext interface was launched in 1993 and quickly replaced by Netscape, which evolved into Firefox.

Netscape and Microsoft created their own proprietary browsers, Netscape Navigator and Microsoft Internet Explorer[8], which they gave away for free (with the former) and bundled in new computer software packages (with the latter). Each interpreted the web code slightly differently, which caused a variety of problems for site designers. The ease of these browsers helped the web take off, with nearly 10 million sites online by 1996.

The early web started a contemporary gold rush, the dotcom boom, with investors pumping money into online ventures they hoped would be profitable. This overheated market collapsed in 2000, taking with it most of the overvalued startups, including PriceLine, Webvan, and Pets.com. While there are few remnants of companies from the dotcom boom and bust today, those that survived are notable: Amazon (1994), eBay (1995), and Google (1998).

eBay, in particular, hinted at the direction the web was headed: engaging users in the creation of a commerce-based community. Founder Pierre Omidyar saw other companies as seeking out "wallets and eyeballs,"[9] a model that imitated the transactional nature of physical-world retail and largely anchored power with companies, not consumers.

Instead, eBay users generated the searchable feed by posting their items for sale and leaving reviews for the sellers they bought from. It was framed as a community. eBay ultimately turned the concept of scarcity on its head, as users dug through their attics and basements to sell treasures and heirlooms that it turned out pretty much everyone had.

During the second wave—the dawn of **Web 1.0**—the end user gained creative access to the network with the assistance of the dynamic duo of personal computers and self-publishing systems, like Netscape Composer and Microsoft FrontPage.

---

6   Of note, the encrypting code was a continuing project of decoders, who finally decrypted *Agrippa*'s code in 2012, two decades after it first encrypted itself!

7   Don't take our word for it; put the phrase "Internet growth summary MIT" into a search engine to learn more (Gray, 1996).

8   Fun fact: Microsoft Internet Explorer launched August 16, 1995, and its "retirement" began June 15, 2022. While that 27-year IE run is impressive, Opera was founded earlier and is still available and supported for the 1% of web users who choose it.

9   " 'Wallets and eyeballs': how eBay turned the internet into a marketplace" by Ben Tarnoff, *The Guardian.*

Online journalism was in its infancy at this point, with newspaper publishers and broadcast producers expressing growing concern as web customers rejected ads and registration and demanded more focus on individual interests. Readers didn't just want to be fed a filtered menu of the day's news; they wanted to be able to select what interested them from the vast buffet of available information.

In the second wave newspapers began making the leap to the web, with most resorting to **"shovelware,"** or simply dumping print content on the web for free. The problem with shovelware is that content from one medium—paper—is dumped onto another—digital—without modification for the medium.

Instead of optimizing stories with hyperlinks, maps, and photos, and publishing them on a 24-hour news cycle, newspapers tended to put their stories on the web in conjunction with the print edition hitting the newsstands. Readers then were expected to pay to read in print what they could read online for free.[10] And there was almost no interactivity.

It wasn't until the late 1990s that the World Wide Web began to show promise as a source of news, information, and entertainment, where information could be posted online far more quickly than it could be published or broadcast.

An unpredictable turning point came when up-and-coming Republican lawyer Kenneth Starr was appointed in 1994 as a special prosecutor to investigate some real estate dealings of Democratic President Bill Clinton.

The investigation's official report was published on the web and people were encouraged to go online and read the Starr Report—in which Clinton was accused of lying during a sworn deposition about his relationship with White House intern Monica Lewinsky.

The findings outlined in the Starr Report led to the December 1998 impeachment of President Clinton on charges of perjury and obstruction of justice. What made the Starr Report so popular on the web? It was probably the warning at the top, which promised "sexually explicit," salacious details of the president's affair, the cigar, the blue dress.

The interest of the public in finding the Starr Report was helped along by the popularity of up-and-coming search engine Google, which had made its debut in mid-1998.

Web users also began to recognize that the medium was literally created to encourage their participation, which eventually led to the success of blogging sites like Blogger and LiveJournal.

The Third Wave: 2001-2007

The third wave of digital media can be categorized as a **BB gun approach,** in that the emphasis shifted to the needs and wants of an individual within a targeted demographic.

Whereas the bazooka had the capacity to take out the large, red barn in one shot, and the shotgun allowed for the targeting of specific parts of the barn, the BB gun analogously takes aim at the chicken scratching for feed in front of the barn. The barn is still there, but it's no longer of interest.

The third wave cemented the web as an excellent source for breaking news. It began in about the fall of 2001, when the world turned to the web as a source not only for information about the 9/11 attacks but also as a place to post their own images, feelings, and news. Other significant developments of the year included Apple's release of both the iPod and iTunes and the debut of Wikipedia.

The third wave of online journalism also brought us **Web 2.0,** which is/was characterized by three significant things:

---

10   Remember that old phrase, "Why buy the cow when you can get the milk for free?" That's part of why legacy media have failed to thrive on the World Wide Web.

- **Personalization**

  Readers now were able to customize news sources and headline feeds that appeared on web browser home pages such as MSN.com, iGoogle, and Yahoo!

  The third wave also introduced open-source news sites like OhmyNews and Wikinews that sought "citizen reporters" to add to the news content. The individual-witnessing model of citizen journalism was a significant challenge to traditional journalism, which channels news through a variety of editorial processes to ensure accuracy and ethics.

- **Participation**

  This marked the departure of the long-held command-and-control world of legacy media. User participation grew exponentially to include blogging, commenting on articles, reading and writing product reviews, generating wiki content, and uploading user-generated photos, audio, and videos.

  Many successful Web 2.0 sites have relied on **crowdsourcing**, or the collaborative generation of information. Wikipedia and Craigslist are notable early examples, with all content generated by users. In a contemporary context, *all* social media sites are crowdsourced, with users both providing and consuming content.

- **Partnership**

  The idea in the third wave was that both the end user and platform services had to give up some power in an effort for collaborative sites to succeed.

  For our purposes, a **Platform Service Provider** (PSP) is a website or app that hosts user content, generally for free. Unlike the **Internet Service Provider** (ISP), which simply provides access to the internet, a PSP allows a user to create and share content. As partners, PSPs offer free services and free consumption of content, while users give up a certain degree of privacy as compensation.

  People became more accepting of advertisements on websites during the third wave and the need to register and/or "approve" user agreements before accessing information and features on individual sites.

  Users also demonstrated a willingness to pay for quality services, whether by providing demographic information on registration or buying month-to-month site subscriptions. In exchange for end-user cooperation, site owners learned to listen very closely to what users wanted.

Some of today's most popular websites were launched as third-wave startups, including Facebook (2004) and YouTube (2005). While Mark Zuckerberg's name is inextricably linked to Facebook's founding (sometimes in not-so-flattering ways), the identity of the young developers who started YouTube, and its origin story, were lost to many when the fledgling video site was sold to Google for $1.65 billion less than two years after its launch.

YouTube's origin story itself is anchored (in part) in a bit of American pop culture drama: the infamous "nipplegate" controversy stirred by the 2004 Super Bowl halftime show.[11] Performers Justin Timberlake and Janet Jackson took the stage to perform "Rock Your Body" live to an audience of roughly 150 million television viewers. At the end of the song, in time with the lyric "I gotta have you naked by the end of this song," part of Jackson was, indeed, naked-ish.

---

11   "Here's how Janet Jackson's infamous 'nipplegate' inspired the creation of YouTube," by Nathan McAlone, *Insider*.

And it was one of those water-cooler moments that *everyone* was talking about the next day. Jawed Karim, a 24-year-old Silicon Valley tech worker, saw an opportunity when he couldn't find the video online.[12] One year later, Karim and collaborators Steve Chen and Chad Hurley had the world's first video sharing site online. Karim, in fact, posted YouTube's first video.[13]

Two influential online community sites, 4chan and Reddit, emerged in the third wave and remain popular today. 4chan (2003) started as an anime-focused forum that allowed anonymous users to post and comment on images, but has evolved into a notorious, opinion-heavy posting site where users contribute to threads on an incredible range of topics. Reddit (2005) is a popular social news website that allows users to create discussion threads around stories that interest them. Users "vote" for stories they like, with topics aggregated on the site, ranked by popularity, and linked to the original source.

The third wave essentially began the trend of people getting their news from social media sites, rather than news sites. It's also when web users realized they could/should have it all and journalists learned that they needed to rethink their role as information gatekeepers. This is where the future for journalism came down to one question: "What do readers want?" Armando Acuña, ombudsman for *The Sacramento Bee*, asked and answered that question in a 2006 column:

> In my view, they want everything, and they want it free.
>
> They want news, and they want it right now and wherever they are.
>
> They want audio and streaming video.
>
> They want events and issues analyzed and explained in-depth.
>
> They want their newspapers to be independent and to be watchdogs of government and the community.
>
> They want to know what is happening in Washington and in Iraq.
>
> They want to know how their local high school football team played Friday night. They want to know about the best night clubs and the newest restaurants in town.
>
> They probably always wanted it all, but couldn't get it until the Internet came along.[14]

The Fourth Wave: 2007-2016

If the first wave of online journalism can be likened to a bazooka that targets as much of the audience as possible, then the fourth wave is akin to **"magnifying-glass-and-ant."**

The big, red barn and the chicken in the barnyard are still there, but they simply don't matter as much anymore to much of the population. The individual holding the magnifying glass—the digital information consumer, for our purposes—has caught sight of a teeming anthill and, after a satisfying kick, is now focusing attention on first this ant, then another, in rapid succession.

In other words, control of the "weapon" had entirely shifted; users were more inclined to engage with and contribute to topics of deep personal interest. As a result, it was increasingly difficult to

---

12 Thanks to the magic of YouTube, you can now find plenty of user-uploaded clips of the 2004 "wardrobe malfunction," along with armchair analysis of whether it was planned or not. We don't mind steering you toward the content, because it is (a) remarkably poor video, and (b) contains far less nudity than most streaming video. That said, by contemporary standards, the song and choreography are notably offensive, for entirely different reasons (which is explained in some depth in other videos).

13 Today the 18-second "Me at the zoo" video has more than 236 million views. It's worth watching, as its composition, quality, brevity, and content make YouTube's ultimate success quite surprising.

14 "Year of changes for newspapers; more coming" by Armando Acuña, *The Sacramento Bee*.

compel citizens to turn away from "The Daily Me" (a very ego-centric creature) and attend to news that has a genuine impact on the community at large and the lives of all.

It was during the fourth wave that paywalls for news sites began to proliferate. Although *The Wall Steet Journal* has had its paywall since 1996, rejecting the idea that news coverage should be free for online readers, most news organizations today have adopted some sort of paywall, usually allowing a limited number of articles per month for free. Some organizations like *The Guardian* and NPR encourage donations from readers and listeners to support the newsgathering and avoid implementation of a paywall.

The fourth wave is a study in cultural disruption, facilitated by the release of Apple's iPhone in summer of 2007. The iPhone was not the first smartphone, but was different from anything else on the market. It combined the qualities of the iPod Touch with the power of a microcomputer, improved the utility of cellphones, and wrapped it all up in a sleek little aluminum-and-glass package. The iPhone was, however, more expensive than any other cellphone on the market ($600, while its closest competitor, the Blackberry, was $200) and required a two-year service contract, available only through AT&T

A Brief History of Mobile Phones[15]

1979—first 1G mobile communication network launched in Tokyo (1983 in U.S.)

1983—the first mobile phone, Motorola's DynaTAC 8000X
   Cost: $4,000
   Weight: 2.5 pounds
   Talk-time battery life: 35 minutes
   Notable features: portable phone, nicknamed "The Brick"
   Network: 1G

1991—first 2G network launched in Finland

1994—the first smartphone, IBM's Simon Personal Communicator
   Cost: $1,100
   Weight: just over a pound
   Talk-time battery life: 1 hour
   Notable features: sent and received faxes and emails, included address book, calendar, clock, notepad

1998—the first internet-connected smartphone, Nokia's Communicator 9000
   Cost: $800
   Weight: just under a pound
   Talk-time battery life: 3 hours, standby 30 hours
   Notable features: all of the Simon, plus web browser, full QWERTY keyboard (allowed typing, rather than triple-tap texting)[16]

2001—first 3G network launched in Japan (2003 in U.S.)

---

15   Adapted from "Smartphone history and evolution" by Meghan Tocci, published by SimpleTexting, August 2019. Specs from PhoneDB, manufacturers, and reviews. Mobile network history adapted from "From 1G to 5G" by Reinhard Haverans, published by BrainBridge Workforce Solutions, May 2021.

16   Triple-tap texting was the realm of the candy bar phone, a block cellphone with hardware numerical keypad on the face and small screen, not a touch screen. To text, you'd use the numerical keypad to type. For example, 1 is ABC, 2 is DEF, 3 is GHI, and so on.

2007—iPhone, 1st gen
Cost: $600
Weight: about 5 ounces
Height: 4.5 inches
Width: 2.4 inches
Depth: 1/2 inch
Display: 3.5"
Camera: 2 megapixels
Talk-time battery life: 8 hours, standby 250 hours
Network: 2G
Notable features: true touchscreen (no stylus needed), haptic touch feedback, rear camera

2008—first 3G iPhone

2008—release of Android operating system 1.0

2008—first Android-based smartphone, T-Mobile G1 / HTC Dream 100
Cost: $179
Weight: about 6 ounces
Talk-time battery life: 5 hours, 5 days on standby
Network: 2G
Notable features: touchscreen + slide-out keyboard, access to the Android Market

2009—first 4G networks launched in Sweden and Norway (2012 in U.S.)

2010—first iPhone with a front-facing camera, iPhone 4
Cost: $4,000
Weight: about 5 ounces
Talk-time battery life: 7 hours, 12+ days on standby
Network: 3G
Notable features: selfies!

2018—first 5G network launched in South Korea (2019 in U.S.)

2022—iPhone 13 Pro Max
Operating system: iOS 15
Cost: $1,100
Weight: about 9 ounces
Height: 6.33 inches
Width: 3.1 inches
Depth: 3/10 inch
Display: 6.7"
Battery life: 18 hours
Network: 5G
Notable features: wide-angle and telephoto cameras, face ID, LiDAR scanner, barometer, accelerometer, 3-axis gyrometer

2022—Samsung S22 Ultra
Operating system: Android 12
Cost: $1,200
Weight: about 9 ounces
Height: 6.43 inches
Width: 3.1 inches
Depth: 3/10 inch
Display: 6.8"
Battery life: 12+ hours of constant use

Network: 5G
Operating system: iOS 15
Notable features: built-in stylus, ultra-wide, wide-angle and 2 telephoto cameras, ultrasonic fingerprint and face recognition

Had Apple been the only company to introduce a smartphone, or had it been even slightly less well designed, the cultural impact of the iPhone might not have been as significant. But Apple did something very clever: it integrated physical world representations—**skeuomorphic design**—into the digital screenscape to help users understand device utility.

iPhone icons *showed* us what we were supposed to do with this shiny new device. We didn't have to wonder what the icons were for, as they looked like their analog counterparts. For example, the notepad app looked like a lined yellow notebook; the calculator buttons looked like, well, calculator buttons; the camera, when taking a picture, appeared as and sounded like an actual shutter; music was represented by album covers you could flip through as you would in the bins at a record shop; and the "newsstand" placed digital publications on a what appeared to be a wooden bookshelf.

Skeuomorphism bridged the gap between the "real" and digital worlds so successfully that smartphones ultimately bumped independent alarm clocks, wristwatches, calendars, notebooks, address books, and photo albums from daily lives. Skeuomorphic design held sway in the iPhone until the release of iPhone Operating System 7 in 2013, which introduced flat design. People had wholly accepted the usefulness of the device, so its utilities no longer needed to look like notebook paper or have 3D buttons. The shift to flat design carried the sleekness of the smartphone's exterior to its digital interior.

The iPhone's cultural impact was not immediate, but gradual, in that at least two other things had to happen for smartphones to become both ubiquitous and indispensable.

First, Apple had to have nimble, less expensive competition, which came with the introduction of Google's Android operating system in fall of 2008. Rather than limiting Android to one smartphone brand, as with Apple's iOS, Android could be used for free by any manufacturer, developer, or consumer.[17]

Second, there had to be room for more than just operating system-specific utilities—native apps—available to individualize these pocket computers. Independently developed apps were integrated into smartphones by 2008, and in 2010 "app" was selected as "word of the year" by the American Dialect Society, citing the ubiquity of Apple's trademarked "There's an app for that" slogan.

By 2011, smartphones were less about phones and more about texting, selfies, video chats, and social media sharing, due largely to the seamless coupling with social media apps, including Facebook, Twitter, Instagram, and Snapchat. This points to another characteristic of the fourth wave: hypercommercialization. As you'll learn in coming chapters, you don't just consume a digital product, you *are* one.

Another significant fourth-wave innovation was the introduction of the Apple Watch in 2015. It took a couple years for it to find its place, but now has numerous competitors. All capitalize on the integration of smartphone and wearable technology, including things like GPS tracking and health monitoring apps.

With the rise of smartphones and proliferation of apps, the web itself became less of a player in the fourth wave. This goes back to the very real distinction between the Internet and the World Wide Web, which is the topic of Chapters 9 and 10, respectively.

---

17 Alphabet does, however, charge manufacturers for integration of Google Mobile Services, such as Gmail, Google Maps, and Google Play.

The Fifth Wave: 2016-Present

At this point, we're a long way from the simplicity of the barn imagined in the first wave. The weapon, to logically follow the preceding analogy, now approximates an all-but-invisible **laser beam** that uses algorithms to profitably isolate individuals within mounds of hyper-personalized interest. While the focus was on the individual in the fourth wave, the fifth wave is characterized by that division pushing us toward ideological colonies that shape and capitalize upon our personal needs, wants, and perspectives.

If we're thinking about the fourth wave akin to a swarming anthill of digital content options, then the fifth wave is closer to a handful of towering termite mounds that now contain the bulk of digital content and commerce. Imagine the tallest towers with labels that read Alphabet, Amazon, Apple, Meta, and Microsoft. There are also smaller towers, representing the plethora of stand-alone sites we routinely turn to for news, information, shopping, and entertainment. Plus, to see inside, you have to *be* inside—a registered, willing participant. You can move from tower to tower effortlessly, but you'll also be tracked when you leave and continually rewarded for returning. The consumer has become the *consumable.*

This points to several fifth-wave themes that will be explored in coming chapters, including:

- Continued consolidation of media ownership and the extraordinary growth of Big Tech;

- The impact of algorithm-selected content, commoditization of privacy, and reality of smartphones as extremely handy surveillance devices; and,

- Our relationship with tech and each other, as digital devices and the internet of things set deep roots in our post-pandemic world.

Before we dig into the past and the roots of all this disruption, we're going to take a short detour to the present to discuss the recent history and profound influence of media ownership in the United States. Onward to Chapter 3.

Summary

Legacy and digital media have fundamental differences with regard to content, audience, and delivery systems. But the distinction between legacy media and digital media—"old" versus "new" media—is inaccurate when considering contemporary news organizations with legacy media roots. Most so-called legacy media have evolved, right along with information technologies, in five general waves: First wave, chatrooms on the internet; second wave, the World Wide Web; third wave, the rise of user-generated content afforded by Web 2.0; fourth wave, smartphones and individualized digital content; and, fifth wave, hyper-personalization, digital surveillance, and the rise of big tech.

Key concepts

Bazooka approach
BB gun approach
Characteristics of Web 2.0
Crowdsourcing
Digital media
False dilemma
Fifth wave of digital media
First wave of digital media

Fourth wave of digital media
Internet Service Provider
Laser beam
Legacy media
Magnifying-glass-and-ant
Platform Service Provider
Second wave of digital media
Shotgun approach
Shovelware
Skeuomorphic design
Third wave of digital media
Web 1.0

References

Aamidor, A., Kuypers, J., & Wiesinger, S. (2013). The evolution of media technologies. In S. Wiesinger (Ed.), *Media smackdown: Deconstructing the news and the future of journalism* (pp. 51–71). New York, NY: Peter Lang.

Acuña, A. (2006, December). Year of changes for newspapers; more coming. *The Sacramento Bee,* p. 31.

Gray, M. (1996). *Internet growth summary.* Retrieved from https://www.mit.edu/people/mkgray/net/internet-growth-summary.html

Haverans, R. (2021, May). From 1G to 5G: A brief history of the evolution of mobile standards. *BrainBridge Workforce Solutions.* Retrieved from https://www.brainbridge.be/en/blog/1g-5g-brief-history-evolution-mobile-standards

Madrigal, A. (2013, September). Beyond the shadows: Apple's iOS 7 is all about the screen. *NPR,* All Tech Considered. Retrieved from https://www.npr.org/sections/alltechconsidered/2013/09/03/218533735/forget-folders-apples-ios-7-is-all-about-the-screen

McAlone, N. (2015, October). Here's how Janet Jackson's infamous "nipplegate" inspired the creation of YouTube. *Insider.* Retrieved from https://www.businessinsider.com/idea-for-youtube-came-from-janet-jackson-nipplegate-2015-10

Pryor, L. (2002, April 18). The third wave of online journalism. *Online Journalism Review.* Retrieved from http://www.ojr.org/ojr/future/1019174689.php

Tarnoff, B. (2022, June). "Wallets and eyeballs": How eBay turned the internet into a marketplace. *The Guardian.* Retrieved from: https://www.theguardian.com/technology/2022/jun/16/wallets-and-eyeballs-how-ebay-turned-the-internet-into-a-marketplace

Tocci, M. (2019, August). Smartphone history and evolution. *SimpleTexting.* Retrieved from https://simpletexting.com/where-have-we-come-since-the-first-smartphone/

CHAPTER 3

Corporate Colonialism: Why Ownership Matters

Figure 3.1. Big Tech has grown into a highly influential part of the global economy, sometimes overtaking government's ability to impose meaningful regulation. Pearls Before Swine © 2021 Stephan Pastis. Reprinted by permission of Andrews McMeel Syndication. All rights reserved.

Takeaway: The physical world is a mix of private and public spaces. Online, it's all private space, and profit-centered corporations with little interest in championing free speech host most digital services, from operating systems, to apps, to the World Wide Web.

Colonialism is the idea of a dominant country establishing a colony within another—usually a developed country with wealth and power, laying claim to parts of a developing country. It's an underlying philosophy that one country knows what another needs, based largely on its own model of governing. Think of Great Britain's expansion into the Americas, Africa, and India. While the empire long ago contracted, its influence still is very apparent in language and governmental structure—in some cases hundreds of years later.

A related concept is **settler colonialism**, which is when the dominant power actually takes up residence in the colony, replacing native or indigenous culture with its own, and exerting sovereignty.

With both of these, there are material consequences, most of which flow because of the relationships of power involved. Self-determination of groups with less power is often sacrificed in the process. In the same way, when you think about the role of media in your culture, it can be clarifying to keep these notions of colonialism in mind.

Corporate colonialism is the idea of corporations taking control of societal institutions that once were considered in the public domain. There are myriad examples of this phenomenon in the United States, from for-profit universities, to corporately owned medical centers, to megachurches.

Corporations began their aggressive colonization of the World Wide Web not long after it became clear there was money to be made. Prior to the web, online services like AOL and CompuServe dominated the market by providing users limited access to the internet, email addresses, and an exclusive space to participate. For that privilege, online service users initially paid *by the hour.*[1]

The growing popularity of the web in the mid-1990s suddenly opened up space for participation and collaboration, and evened the playing field. Users still needed internet service, which quickly evolved to unlimited-use plans, but they could choose to participate where they wanted on the web.

Free Space = Free Speech, Right?

While corporate colonialism in our physical communities has largely economic effects, the same phenomenon online has the capacity to challenge core constitutional freedoms. When American citizens are in public spaces or private homes, they have relatively broad constitutional rights to speak freely without fear of government reprisal.[2]

You can stand on a downtown street corner with a sign that says, "Honk if you love free puppies!" (which is a surprisingly controversial stance). As long as you don't impede traffic, you have the right to be there and share your views on war, the president, gun control, abortion, religion, education, anarchy, big business, or whatever. You even have the rights to do this as a group, or as it says in the First Amendment, the right peaceably to assemble.

In fact, if your perspective turns out to be particularly unpopular and an unfriendly crowd forms around you, the police are obligated by the First Amendment to protect you. If a person in that crowd throws a water bottle at your head, that person could be arrested.

---

1   Yes really. Subscriptions were generally for about 10 hours per month and anything beyond that got expensive fast.
2   The First Amendment protects Americans from *government* abridgment of free speech, among other things.

Think about doing the same thing at, say, Walmart or Target. Bring your signs and claim your space inside the automatic doors, where you can share your views with every person who enters. You'd be removed and likely arrested for trespassing and disturbing the peace.[3]

There are public spaces where we may freely speak, and there are private spaces where those who own the space may freely limit speech.

Now, let's apply that to the World Wide Web. Is there public space? Is there free speech? Of course there is! Whatever site you sign up for provides your own little bit of space to do with what you choose—Facebook wall, YouTube channel, Pinterest board, Twitter feed, and so on. Just like a rented apartment, you can decorate your space and you can invite in whomever you choose, but you can't change the basic layout.

Also, like an apartment, *it's not really your space*—it belongs to whatever company owns the website or app. The bulk of these services are free, while an apartment is not free; money must change hands to guarantee your rights within that space.

If use of a website or app can be compared to renting an apartment, is that *your* personal space? Is it public space, with a guarantee of free speech? The answer to both is simple: No.

There are millions of spaces to post personal opinions, photos, and videos, but we don't really own or rent those spaces. We are invited guests who are welcome to stay as long as we abide by house rules.

In the parasitic realm of corporately owned **Platform Service Providers**—companies that provide you space to participate in the digital conversation—violating the terms of a user agreement is akin to vomiting in your host's pool. If your post offends someone or you use copyrighted material one time too often, you—the virtual you—can be removed and not welcomed back.

The founders of the World Wide Web envisioned it to be something like an infinite public square— a boundless territory where people could share their ideas and opinions across space and time. They would be free to collaborate, create, and innovate.

Today's web is more like an endless expanse of big-box stores—one massive private space after another. It costs you nothing to walk through the door, but there is a tremendous social cost to these private spaces being there. In the physical world it may be the erosion of downtowns; on the web, it's the erosion of your privacy.

Purpose as a Means For Profit

Corporations typically are structured with an eye to economic value. They want to make money, grow, and make more money. Every service acquired, offered, and operated is intended to generate a healthy fiscal return on investment, or the corporate organism cannot survive. That's what literary theorist Kenneth Burke[4] referred to as **Agency**. On the flip side is **Purpose**, which is more closely aligned with the end product and its ability to induce attitudes or further action.

Agency manifests in the corporation maintaining its right to offer platform services on an international scale, while attempting to distance itself from the rights and responsibilities afforded to citizens of individual countries. In short, Alphabet (Google) wants its search engine, workspace, email system, cloud service, video-sharing service, and so on, to be the choice of every digital media user in every country.

This preoccupation with value and function does not preclude Purpose, but it may embrace it largely as a means toward an end. Good service begets good revenue. Purpose, then, can be seen as an ideal that may be fostered organizationally as long as it benefits the financial bottom line, meets corporate expectations for future growth, and boosts stock prices.

---

3   You might think of the "protests" staged by people who refused to abide by the masking rules in these private spaces. Frequently they were escorted out, often with the presence of law enforcement.

4   Burke (1969).

For these needs to be met in the physical world, corporations must be responsive to a wide range of publics—neighbors, employees, consumers, government regulators. Digital tech companies still have more of a Wild West, take-it-or-leave-it approach to social responsibility. The company wants you to adopt, adapt to, and keep coming back for what it offers. The potential for negative environmental or social outcomes is not necessarily part of the equation, as the services it provides are less tangible than physical products in the physical world.

Yet platform services are powered by massive data centers that suck a tremendous amount of electricity. The data centers exist to meet individual and organizational demand for platform services, not to exemplify social responsibility or facilitate common good.

Where Is Common Good in the Digital World?

Common good is the idea that there are things in society that benefit all. In the United States, there are numerous examples of partnerships between the people and the government that are driven by a desire for shared services and stable social structure. We pay income, property, fuel, luxury, and sin taxes,[5] and the government puts that money toward organizations that preserve democracy and protect the things we have determined to be in our best common interests, whether it be the military, social services, fire departments, public libraries, or the public school system.

The U.S. interstate system is a tangible example of the notion of common good. If you want to drive coast to coast, it's likely that you'll pick a route that at some point joins Interstate 40, 80, or 90, which are the most-direct, best-maintained routes. While relatively few Americans use the nearly 50,000 miles of interstate system every day, most benefit from its existence and all taxpayers contribute to its maintenance in one way or another.

When we buy products that are trucked into our communities, we're helping pay the cost of the interstate by absorbing user fees, fuel taxes, and tolls paid by the trucking companies for use of any part of the national highway system. In addition, every time motorists buy a gallon of unleaded gasoline or diesel fuel, they are paying up to a quarter per gallon to help fund the interstate system.

How Is Digital Different?

We don't pay taxes to fund the information superhighway, but the World Wide Web arguably is a form of common good, in that it offers access to seemingly limitless free information. But the notion of common good is that there is a shared sense of contribution—that we understand that each person paying a little bit benefits everyone.

When digital media users pay, they typically are paying for *their* internet/phone connection, for the level of speed/data *they* prefer, to power the devices *they* have purchased. That money goes to the Internet Service Provider and the device seller. Sure, the government gets its share in the form of sales and use taxes, but the money is not necessarily earmarked to fund improvements to the country's information infrastructure. That's more typically the realm of private enterprise in the United States. You can see how some of this is transacted if you look closely at telephone, wireless, cable, and internet service bills, where you will find a miscellany of regulatory recovery fees and surcharges.[6]

---

5   Luxury taxes are assessed on things like expensive jewelry, furs, cars, private planes, and boats. The idea is that these things are not essential to everyday life and that only people of a certain income can afford them. Sin taxes are assessed on things that are deemed to be driven by individual choice and may be physically and socially destructive, like cigarettes, alcohol, and gambling. By the time you read this, fast food, products containing high fructose corn syrup, and sugared sodas may all be subject to sin taxes. Just thinking out loud here: What are the odds that purely entertainment websites—from gossip to porn—could someday be subject to a sin tax?

6   You can find a lot of information about this confusing world of fees by searching online for "Broadband bill autopsy: The origin and purpose of every single fee" by Tyler Cooper, BroadbandNow.

Online common good also is complicated by a system that is inherently driven by user choice—from hardware, like computers and mobile devices, to software, such as operating systems, apps, browsers, and search engines.

There are thousands of brands, devices, and programs that are used in highly individualized combinations to give each user a slightly different experience. No two visits to the web are the same. We can fully intend to be productive and still end up watching giraffe videos or scanning celebrity gossip for half a morning.

Things representative of the common good tend to develop in a methodical way, necessarily drawing in support from various constituencies along the way. Digital information tends to spread in a random, viral manner, from human to human, interface to interface. We find things via each other, and what we find tends to remain in the orbit we've created via our online and mobile networks.

Something that helps a society develop evenly and purposefully is common good; something with an innate capacity to distract us from work, family, and friends displaces common good with an intoxicating potion of highly individualized needs and wants.

How Do Platform Service Providers Make Money If Everything Is Free?

The legacy media business model involved shooting necessarily limited information out to an undifferentiated—mass—audience. When you picked up your newspaper from the porch or tuned in to the 6 o'clock news, you were getting the same thing that everyone else did. The **legacy media business model** looked—and continues to look—like this:

- Content producer (journalists)

- Content distributor (legacy media outlets)

- Content consumer (the public)

- Largely funded by generalized advertising from businesses within the geographic proximity of the publication/broadcast, or purchases like buying a movie ticket or a magazine or newspaper at a news stand.

With this model, the newspaper or broadcast company sells advertising and resulting revenues fund the product and support the organization. Those who produce content, both news and ad copy, are paid for their efforts. When combined with the purchase of individual goods, this represents can be considered dual revenue streams.[7]

This is in stark contrast to the Platform Service Provider business model, where *we* are content producer and consumer. The service provider is in the middle, functioning with the Agency-centric goal of getting as many people as possible to engage with a product and keep coming back to it. The **platform service business model** looks like this:

- Content producer (us)

- Content distributor (Platform Service Providers)

- Content consumer (us)

---

7   Havens and Lotz (2017), p. 119.

– Largely funded by individualized advertising from businesses that sell products of interest to individuals (based on a complex formula that considers the things they view online, where they are, the content of their social media postings, and who they engage with)

While advertising is the primary income source in both models, the way companies make money from the advertising is completely different. With the legacy mass media model, advertising is sold based on newspaper circulation—how many hands a typical edition passes through—or ratings for particular television shows. The ad is sold, published/aired, and ultimately helps pay for both creation of content and its distribution.

A digital version of the same content also is monetized through advertising, but digital compensation is far more complex. Digital revenues are not about content; they're about *distribution* of content and *control* over revenue sources.

Notably, Alphabet and Meta acquired some of the largest digital advertising companies, with the two controlling 68% of a $252 billion digital ad market in 2019,[8] a virtual duopoly. Of that total, Alphabet controlled 41% and Facebook (which was still Facebook) took in 27%. Interestingly enough, by 2019, Chinese companies controlled the next largest amount of digital advertising at 22%.

By contrast, in 2015 the digital ad industry generated about $60 billion, with 41% of that revenue still going to legacy media companies. Google (which was still Google) controlled about 16% of that market and Facebook took in 30%.[9]

Digital advertising requires a combination of meticulously tracked **view-throughs** and **click-throughs.** A view-through involves someone landing on a particular page and the advertiser paying for the opportunity to get attention for an ad. Click-throughs involve more than a potential view; the Web user has to click on an ad link and follow it out of the site.

Instead of featuring only ads sold by employees and relevant to the local market, legacy media companies enter into agreements with the tech companies who own the bulk of the digital advertising business. With digital advertising distribution concentrated with a few large tech companies and news stories distributed via social media channels, news organizations lost 62% of ad and circulation income between 2000 and 2018.[10]

Ads are targeted to users, regardless of the site, based on specific demographic parameters set by the advertiser. If it feels like an ad is following you around as you move from site to site, it's because it is. Alphabet and Meta effectively track you from site to site, platform to platform, continually placing personally relevant ad content in your path.

Believe it or not, when personalized ads first started appearing on Facebook pages, people were disturbed—not so much by the personalized ads that were on target but by the misses. If you just posted news of your engagement to be married, it makes some sense that you'd get ads for honeymoon destinations. But what about the weight loss ad below it or the one about middle-age incontinence?

Solving that problem has changed the way we interact with both the web and web advertising. Why wouldn't you want personalized ads? The closer a site can target your interests, the better the user experience.

But the platform services business model demonstrates a key problem with the parasitic relationship between Platform Service Providers and web content providers. Neither reporter nor news organization is paid for original news content circulated on social media. While there are ads on social

---

8   "Net US digital ad revenues, by company, 2015–2020." *Insider* Intelligence.
9   Ibid.
10   "Fast facts about the newspaper industry's financial struggles as McClatchy files for bankruptcy" by Elizabeth Grieco, Pew Research Center.

media sites, those ads benefit the Platform Service Provider. If a news organization can't get you to the "mothership"—its home website—it's not making the money needed to pay for content creation.

Similarly, if a popular reporter builds a following on social media, web users may be *following that person* and never engaging with the source organization. One example is the development of Substack, a platform where writers can assemble their own core of subscribers, or can distribute their material free if they wish. And, again, it's the Platform Service Provider that makes money from the content, not the reporter or news organization.

Facebook, Instagram, Twitter, YouTube, TikTok, Substack, and Google-as-search-engine exist because *you* post. And they make a lot of money from your ubiquitous engagement with their "free" services and uncompensated work as a human cog in their digitized machines.

This relationship requires corporations to constantly seek innovation that quickly reaches as many people as possible. And it's not enough to have an audience; a successful platform service requires *mass adoption* in an extraordinarily competitive environment. Services that don't hit critical mass quickly enough tend to have a very short digital life.

Crowding Out the Competition

It's important to recognize that consolidation of ownership isn't new. The absolute power of Alphabet, Amazon, Apple, Meta, and Microsoft—which we'll collectively refer to from this point forward as the **Frightful Five**—is the change we need to pay attention to. Farhad Manjoo coined that very apt term in 2017,[11] pointing to how our attraction to convenience makes us complicit in the domination of these companies.

Manjoo was particularly concerned about the ease with which Amazon has channeled a significant degree of commerce through a single online source. He noted,

> you can decide to opt out; you can drive to Target, your life won't end if you don't patronize Amazon. But if it's not Amazon for you, it'll be another of the five. Or, more likely, it already is. It's too late to escape.

And, if anything, the pandemic cemented our relationship with the Frightful Five and other tech services that kept us entertained, supplied, and connected as we rapidly transitioned to work- and school-from-home.

The history of media ownership in the U.S. may well explain why, culturally, we've largely allowed the Frightful Five to grow into the corporate behemoths they are today.

Consolidation of newspaper ownership in the United States dates back to the late 1800s, when Edward W. Scripps discovered the advantages of proclaiming news objectivity in the heyday of the partisan press. He and his brothers started buying out the competition at a time when many newspapers existed.[12]

That lead was followed in the early 1900s by Frank Gannett, who began acquiring newspapers in the area surrounding his first property in Elmira, New York. Scripps and Gannett each were so successful that the companies they began endure today as among the largest newspaper groups in the nation.

Still, prior to World War II, most communities had at least one and often two or three family-owned newspapers, engaged in competition for both advertising dollars and the news. Between 1946 and 1999, however, daily newspapers transformed from 75% local family ownership to 73% corporate ownership.[13]

---

11 "Tech's Frightful Five: They've got us," by Farhad Manjoo, *The New York Times*.
12 Kaplan (2002) and Compaine and Gomery (2000).
13 Bagdikian (2000).

For example, in 1995 Disney acquired Capital Cities/ABC, a media company that owned a few dozen newspapers, radio stations, and ABC television affiliates for $19 billion. Disney wanted control of ABC, but had no interest in the community newspapers. They were quickly sold.

In 1983, a media scholar named Ben Bagdikian[14] claimed that 50 companies control what Americans read and watch.[15] In 2000, he updated that to *six* media companies that dominate all of American mass media.[16] In 2004, Bagdikian said that *five men* dominate the world's mass media, including the CEO of Disney and the owner of News Corp.[17]

Hulu: A Case Study in Consolidation of Ownership

Hulu today is one of a growing number of subscription streaming video services. It not only was one of the earliest, but is a spot-on illustration of the evolution of media ownership.

Hulu was created to compete with YouTube, in response to NBC-owned shorts from "Saturday Night Live" going Viral. One of the first was "Lazy Sunday: Chronic of Narnia," a clip that went viral December 17, 2005. It also prompted NBC to demand that the user-posted clip be taken down, which users took as a challenge to keep uploading copies.

A year later, another SNL short, "Dick in a Box,"[18] went viral and it was clear that NBC had recognized the potential of an exclusive location for NBC's film and TV library. Hulu, then, is the moderately successful offspring of legacy media titans in their bid to compete with digital media. Here's a brief history of that evolution:

1912—Universal Studios founded in Hollywood

1926—National Broadcasting Company (NBC) founded by Radio Corporation of America (RCA)

1964—Universal Studios theme park founded

1981—Seagram's, a very old Canadian company that was once the largest alcohol beverage company in the world, acquires Universal Studios and Theme Parks

1986—General Electric, one of the most powerful U.S. companies, acquires NBC

1997—A very old French company rebrands itself Vivendi and begins offering telecom services

2000—Vivendi Universal is created in merger among Vivendi, Seagram's, and Universal

2004—GE merges with Vivendi Universal to create NBC Universal

2007—Hulu is created, owned by

- – NBC Universal (60%)
- – News Corp / 21st Century Fox (30%)

---

14   Ben Bagdikian was a former journalist, credited with helping get the Pentagon Papers published in The Washington Post in 1971. He became an ardent and influential media critic and was a professor at UC Berkley until his retirement in 1990. He died in 2016, at the age of 96.

15   Bagdikian (1983).

16   Bagdikian (2000), op. cit.

17   Bagdikian (2004).

18   Fun fact: "Dick in a Box," which starred Justin Timberlake and Andy Samberg, won a Primetime Emmy Award for Outstanding Original Music in 2007.

- Providence Equity Partners, a private investment firm (10%)

2011—Comcast buys controlling interest in NBC Universal

2013—Comcast becomes sole owner of NBC Universal

2018—Comcast splits its 60% Hulu stake with Disney, leaving Hulu owned by

- Comcast (30%),
- The Walt Disney Company (30%),
- News Corp / 21st Century Fox (30%),
- AT&T / Warner Media (bought out 10% stake of Providence Equity Partners)

2019—Hulu's "final" ownership takes shape

- AT&T's 10% stake in Hulu is bought out by Disney & Comcast
- Disney buys 21st Century Fox for 67% ownership;
- Disney buys out Comcast for full ownership

Perspective:

A partial list of what Comcast owns:

- NBC / Peacock
- Universal Studios
- Dreamworks
- MSNBC, USA Network, SyFy, E!
- And it's the largest cable/internet provider in U.S.

A partial list of what Disney owns:

- ABC
- 21st Century Fox
- Hulu
- Marvel Entertainment
- Lucasfilm
- Pixar
- ESPN, National Geographic, A&E, History Channel
- And a few theme parks …

Every single one of the companies listed above were once large independent companies in their own right that were acquired by Comcast and Disney along the way. The *combined* value of these two enormous media companies is *half* the value of Meta, which, at this writing, is the least profitable of the Frightful Five.

What Makes These Companies So Powerful?

It's not hard to imagine what Bagdikian would have to say about media ownership today, when five technology companies—that contribute little to no original content—control the attention span of the American public, distribution of a colossal amount of content, and virtually all digital advertising.

Individually, each of the Frightful Five are among the wealthiest companies on the planet. Each is publicly traded, with millions of shareholders. In 2018, Apple was the first trillion-dollar American company, followed in succession by Microsoft, Alphabet, Amazon, and Meta.

In early 2022, Apple was the first company in the world to top a valuation of $3 *trillion*. That's three *million million*. To put that in perspective, a journalist framed it this way:

> Combine Walmart, Disney, Netflix, Nike, Exxon Mobil, Coca-Cola, Comcast, Morgan Stanley, McDonald's, AT&T, Goldman Sachs, Boeing, IBM and Ford. Apple is still worth more.[19]

Collectively, the Frightful Five have acquired literally hundreds of other tech companies and innovations, and spent hundreds of billions of dollars doing so.

Inventions that pose promise or competition, are acquired by these resource-rich mega-companies, merged into existing products, evolved into branded, proprietary products, and/or purchased for their potential with future innovations.

While Facebook and YouTube may have been created by ambitious 20-somethings during the third wave, the opportunities they had less than two decades ago are effectively blocked today by the mature companies they created.[20]

For example, Palmer Luckey was 20 when he founded Oculus VR in 2012—just slightly older than Mark Zuckerberg was when he created Facebook predecessor FaceMash in 2003. Luckey was 22 when Facebook purchased Oculus VR for $2 billion in cash and stock options in 2014 and welcomed him as one of the company's many developers.

While tech entrepreneurs like Luckey often end up very, very wealthy, they also tend to become dispensable cogs in the Big Tech machine. The Oculus Rift, largely designed by Luckey, was released by Facebook in early 2016. Just over a year later, Luckey was fired by Facebook. After a very successful run of Oculus headsets, the brand was quietly eradicated by Facebook when it rebranded itself as Meta in 2021.

Below are partial lists[21] of notable purchases by each of the Frightful Five. Take a quick look and see what companies/products you recognize:

Alphabet (Google)			
Acquisition	**Year**	**Cost**	**Tech / Talent**
YouTube	2006	$1.65 billion	Video sharing
DoubleClick	2007	$3.1 billion	Online advertising
Motorola Mobility	2011	$12.5 billion	Smartphone tech
Nest Labs	2014	$3.2 billion	Home automation
HTC	2017	$1.1 billion	Smartphone tech
Fitbit	2021	$2.1 billion	Biotech wearables

19   "Apple becomes first company to hit $3 trillion market value," by Jack Nicas, *The New York Times*.
20   If you're ever looking for a good definition of irony, this may well be it.
21   To see more complete lists of what the Frightful Five own, go to Wikipedia and search for "List of mergers and acquisitions by …" You'll see some information about what tech the acquired companies specialized in, as well as what existing products the tech was integrated into. If you're curious about the individual acquisitions, you can search for them separately in The Google.

Amazon

Acquisition	Year	Cost	Tech / Talent
Audible	2008	$300 million	Audiobook service
Zappos	2009	$1.2 billion	Shoes, clothing
Twitch	2014	$970 million	Interactive livestream
Whole Foods	2017	$13.7 billion	Grocery store chain
Ring	2018	$839 million	Home security
Zoox	2020	$1.2 billion	Autonomous cars
MGM Studios	2021	$8.45 billion	Film & TV

Apple

Acquisition	Year	Cost	Tech / Talent
Soundjam	2000	unknown	Foundation of iTunes
Siri	2010	$200+ million	Voice assistant
Beats Electronics	2014	$3 billion	Headphones, streaming
Shazam	2018	$400 million	Music, image recognition
Intel modem div.	2019	$1 billion	Smartphone modems

Meta (Facebook)

Acquisition	Year	Cost	Tech / Talent
ConnectU	2008	$31 million	Facebook competition
FriendFeed	2009	$47.5 million	Facebook competition
Friendster	2010	$40 million	Facebook competition
Instagram	2012	$1 billion	Social media
WhatsApp	2014	$19 billion	Secure messaging
Oculus VR	2014	$2 billion	VR headsets
Giphy	2020	$400 million	Gif creation, distribution

Microsoft

Acquisition	Year	Cost	Tech / Talent
Hotmail	1997	$500 million	Email system
aQuantive	2007	$6.33 billion	Digital marketing
Fast Search	2009	$1.2 billion	Web search engine
Skype	2011	$8.5 billion	Teleconferencing
Nokia	2013	$7.2 billion	Mobile, smartphones
Mojang Studios	2014	$2.5 billion	Video Games (Minecraft)
LinkedIn	2016	$26.2 billion	Professional networking
GitHub	2018	$7.5 billion	Collaborative development
ZeniMax Media	2020	$8.1 billion	Video games
Nuance Com	2021	$19.7 billion	Speech synthesis/recognition
Activision Blizzard	2022	$68.7 billion	Video Games (Call of Duty, Candy Crush)

Along with the actual tech acquisitions, the companies also routinely acquire "talent"—the developers and designers employed by the selling company—who often continue to work on projects related to their previous work. For example, Apple purchased Emagic in 2002 and its sound engineer went on to create Garage Band, which still is bundled with Mac software more than two decades later.

How does this systemic acquisition happen without the attention of governmental regulators? Good question. As antitrust law professor William Kovacic explained,

> A very powerful criticism of U.S. antitrust enforcement, especially over the past 20 years, is that the failure of the agencies to bear down on smaller acquisitions, including acquisitions of—at the time—small companies like Instagram, allowed the preeminent digital firms of today to acquire unassailable positions of dominance.[22]

By the way, antitrust is a very important concept when we think about capitalism, competition, and private companies. Antitrust laws are designed to limit the power of any particular player in a market in order to preserve competition. They are also supposed to stop a set of players in a market from becoming a cartel, or to collude and fix prices, or to form a monopoly, grounded in the Sherman Antitrust Act of 1890 (!). Since then, and at the same time, these players are often wealthy enough to hire lobbyists to work against antitrust actions.

Why Did Google and Facebook Change Their Names?

Alphabet and Meta have considerable influence over huge parts of our daily lives. You likely still think of them as Google and Facebook, but each suite of companies has been renamed to better reflect the vast array of products they own and operate.

The recasting of Google as Alphabet took place in 2015, with Alphabet essentially created to be an umbrella company with separate divisions encompassing Google products, including YouTube and Android; development of acquisitions, including Nest and Fitbit; and an entire innovation division called X, known as the moonshot factory.

The pivot to the name Alphabet was a bit of a puzzle, as (a) it was an exceedingly common, strikingly unrelated term; and, (b) other companies already owned nearly every variation of the web domain and stock market trading names. As a result, typing "Alphabet" into a search engine won't get you to Google's parent company, and if you want to look at the stock price, it remains GOOG or GOOGL.

Alphabet's website is abc.xyz, which, as of this writing, opens to a mostly blank page with "G is for Google" and a short note from founder Larry Page that expands to a very long letter when you click "more." Beyond an investor relations archive, there's literally nothing on the website that links Alphabet to the plethora of companies it owns.

So why did Google and Facebook change their names? It was time. At least that was the official story from each. Yet neither engaged in any immediate corporate restructuring—the divisions that existed afterward, existed before.

But there are underlying reputational reasons why both companies might have sought rebrands that go beyond the sheer scope of technologies they direct the evolution of. Assessments of both name changes largely have focused on company value, stockholder interest, and perceptions of cultural influence.

The Alphabet rename, for example, helped clarify for investors the range of "products" the company offers and was also an attempt to deflecting focus from Google-as-search-engine. This sort of makes sense, given that Google was born in 1998 and has evolved into something far greater than that single search function.

---

22 "Tech giants quietly buy up dozens of companies a year. Regulators are finally noticing" by Gerrit De Vynck & Cat Zakrzewski, *The Washington Post*.

Co-founder Larry Page explained the shift as avoiding perceptions that Google is a settled, conventional company.[23] This name-change rationale was validated and questioned in news coverage.

Yes, Google had grown beyond that single profitable product, so the shift better reflected what Google actually does. As economic policy journalist Matt O'Brien explained,

> It's still a search company that uses its monopoly profits to subsidize speculative science projects and invest in startups. Google was a conglomerate before, and it's still one now, just with a corporate structure that better reflects that fact.[24]

Others pointed out that Google had taken early image hits from those concerned about its growth into areas that compromise user privacy, so the name change allowed a bit of strategic ambiguity. As tech journalist Dave Smith noted, "they're now owned by Alphabet, an innocent-sounding company you've probably never heard of."[25]

The renaming of Facebook to Meta, which was announced in late October 2021, was accompanied by an official rationale very similar to that of Google: A desire to better reflect the company's present and future. Zuckerberg noted that the Facebook brand "is so tightly linked to one product that it can't possibly represent everything we're doing today."[26]

The Facebook rebrand appeared a bit more calculated, however, and was met with corresponding skepticism.

At that point, founder Mark Zuckerberg had landed in front of Congress more than once to testify about the company's role in eroding privacy and democracy. Facebook's amplification of election and pandemic misinformation was on the brink of legendary, and a whistleblower had just weeks earlier testified that the company's approach to social media was one that put profit over user safety. Perhaps worse (for Facebook), its social media platforms had been all but abandoned by younger audiences, in favor of TikTok.

Zuckerberg took the stage at Facebook's Connect conference to announce the company's growing focus on the **metaverse**, an immersive digital environment where people can do most of the things they do in the physical world, but without physical world limitations, such as time and space.

The Meta pivot was not only necessary from a public relations perspective, but also reflected the clear direction Zuckerberg had already begun to take his company. Tech journalist Kevin Roose explained the stakes:

> If it works, Mr. Zuckerberg's metaverse would usher in a new era of dominance—one that would extend Facebook's influence to entirely new types of culture, communication and commerce. And if it doesn't, it will be remembered as a desperate, costly attempt to give a futuristic face-lift to a geriatric social network while steering attention away from pressing societal problems. Either possibility is worth taking seriously.[27]

Future reader, you may already know the answer.

---

23   "G is for Google" by Larry Page. The name, by the way, was an offshoot of the term *Googol*, which a term invented by a mathematician's son to describe one followed by a hundred zeros.
24   "The real reason that Google changed its name" by Matt O'Brien, *The Washington Post*.
25   "One year later, nobody knows what Alphabet is—and that's a godsend for Google's public image problems" by Dave Smith, *Insider*.
26   "Founder's Letter" by Mark Zuckerberg, Meta.
27   "The Metaverse is Mark Zuckerberg's escape hatch" by Kevin Roose, *The New York Times*.

We're Part of the Problem

As Farhad Manjoo noted, we are complicit participants in our domination by Big Tech. We *like* what these companies have to offer. Microsoft hooked the nation by bundling its Office suite with all new PCs. Steve Jobs branded Apple as a company that sells not only really good technology, but *style*.

Even if we *know* that we'd be better off with less social media and technology in our lives, there are five basic **adoption factors** that keep us coming back:

1. **Fear of missing out.**
   You don't want to be the only person in your social group who isn't on social media. You don't want to be the person in the group chat with a green text bubble.[28] That's also known as herd mentality.[29]

2. **Fulfillment of need.**
   Does this thing meet some information or entertainment need that isn't being met in any other way? How do we feel when we engage?

3. **Ease of use.**
   Can't find what you're looking for? Frustrated that something doesn't work? Our egos are not big fans of feeling slow, behind, lost. We like stuff that works predictably every time. If a new app or social media site is difficult to navigate, we have less difficulty rejecting it.

4. **Reliability.**
   Digital media users aren't very patient. And they don't have to be. If something works sporadically or crashes frequently, they're moving on. If it's a hot enough innovation, something better will rapidly replace it.

5. **Comfort.**
   Most humans innately abhor chaos. Even people who fancy themselves as techies can be exhausted by the endless evolution of platform services. There's no way we can keep up, so it's easier to pick a service that works and stick with it.

Each of the preceding factors is either underpinned or can be overwhelmed by **Baby Duck Syndrome**, which refers to the propensity of baby ducks to imprint on the first animal they encounter. From that point on, the baby duck will devotedly follow whatever duck, goose, dog, sheep, or human it believes to be its "mother"—regardless of what the creature does to convince the baby duck otherwise.

Baby Duck Syndrome emerges when a person believes that whatever computer, mobile device, software, browser, search engine, or social media site is the very essence of things of that kind for all related purposes.

People sometimes imprint upon whatever technology they first learned about, were trained with, or had any degree of success with and will continue to use and defend it even as time, technology, and information systems march onward.

For example, if the first smartphone you use and/or buy is an Apple iPhone, then that specific type of mobile device—one of two major operating systems and many brands on the market—may come to represent mobile devices in general. Your vocabulary and perspective on smartphones thus may reflect

---

28   Blue text bubble on iPhone = iPhone. Green text bubble on iPhone = Android. Apple literally flags people in group chats who don't have the more expensive cool stuff.

29   Herd mentality happens when we feel the need to do something just because everyone else is. When someone exclaims, with a look of horror, "What do you mean *you're not on Instagram*?" or "You're *still using Safari* as your browser?," We know we need to be on Instagram and download Chrome. Sometimes we can't help it. Being different, alone, and vulnerable can be dangerous.

only the experience of an iPhone user because that is the vocabulary and perspective the choice itself provides.

One of the symptoms of Baby Duck Syndrome is that the person *qua* duck tends to categorize everything in terms of iPhone and not iPhone—the quintessential thing, not the quintessential thing.

Baby Duck Syndrome can result in **consumer inertia** that benefits established platform service companies and can destroy up-and-coming ones. Inertia is a force that keeps something going in one direction. Consumer inertia, then, would be choosing to stick with one product or company because you're satisfied and/or it's easier than switching.

Did you know Microsoft has a search engine? It's called Bing and it's been around since 2009. If you've never heard of Bing, it's likely because you're not a gamer[30] and the only search engine you've ever used is Google.

Alphabet has capitalized on the ubiquitous use of Google-the-search-engine and Gmail to garner comfortable acceptance of Chrome and Google Workspace (Drive, Docs, Sheets, Slides, etc.).

If you own a Mac, you may be more likely to use Google Workspace than Microsoft Office, as the latter is not bundled with Apple computers. That's because Apple wants you to use its iWork programs (Pages, Numbers, Keynote), which have proven to be far less popular than either Office or Workspace.

If you have had nothing but good experiences with a particular platform service brand, you are more likely to trust in and try its new product releases. Alphabet, Apple, and Facebook may be notorious for privacy breaches, but each has acquired hundreds of millions of users and successfully sold missteps as efforts to improve user experience.

Or you may want a different social space than your parents. That might be one of 17 reasons why you never really wanted to have much to do with Facebook.[31] The question is, will the same thing happen to the next generation when they want their own space, and not your Instagram and TikTok?

A new platform service not only requires the user to learn new tasks but also to trust that the company, the site, and your content won't just disappear one day. In short, the devil you know may be better than the one that just appeared.

Next up: a little perspective on how the technology itself influences our response to it, which is the subject of Chapter 4.

Summary

Increasing consolidation of media ownership has brought forth a new generation of media barons, whose companies largely don't produce much in the way of original content. Instead, these global conglomerates are monetizing our attention, as their revenues, power, and influence grow. Does personal space exist in any meaningful way on the web or apps? An answer: Yes—I can post whatever I want. The correct answer: No. There is no true public space in cyberspace. There are millions of spaces to post personal opinions, photos, and video, but the user does not own those spaces. They are borrowed from site owners—Platform Service Providers—in exchange for our privacy. If you post a video on YouTube or a photo on Instagram, you exist in that space at the whim of the host.

Key concepts

Adoption factors
Agency
Baby Duck Syndrome

---

30   Bing comes bundled as the default search engine for the Xbox.
31   Look for this article online to discover the other 16 reasons: "17 reasons why the kids don't like Facebook anymore" by Katla McGlynn, *Huffington Post*.

Click-throughs
Colonialism
Common good
Consumer inertia
Corporate colonialism
Frightful Five
Legacy media business model
Metaverse
Platform service business model
Platform Service Providers
Purpose
Settler Colonialism
View-throughs

References

Bagdikian, B. H. (1983). *The media monopoly*. Boston, MA: Beacon Press.

Bagdikian, B. H. (2000). *The media monopoly* (6th ed.). Boston, MA.: Beacon Press.

Bagdikian, B. (2004). *The new media monopoly*. Boston, MA: Beacon Press.

Burke, K. (1969). *A grammar of motives*. Berkeley: University of California Press.

Compaine, B. M. & Gomery, D. (2000). *Who owns the media?: Competition and concentration in the mass media industry* (3rd ed.). Mahwah, NJ.: L. Erlbaum Associates.

Cooper, T. (2021, October, 26). Broadband bill autopsy: The origin and purpose of every single fee. *BroadbandNow*. Retrieved from https://broadbandnow.com/report/broadband-bill-autopsy-origin-purpose-every-single-fee/

De Vynck, G. & Zakrzewsi, C. (2021, September 22). Tech giants quietly buy up dozens of companies a year. Regulators are finally noticing. *The Washington Post*. Retrieved from https://www.washingtonpost.com/technology/2021/09/20/secret-tech-acquisitions-ftc/

Grieco, E. (2020, February 14). Fast facts about the newspaper industry's financial struggles as McClatchy files for bankruptcy. *Pew Research Center*. Retrieved from https://www.pewresearch.org/fact-tank/2020/02/14/fast-facts-about-the-newspaper-industrys-financial-struggles

Havens, T. & Lotz, A. (2017). *Understanding media industries*. New York: Oxford.

Kaplan, R. L. (2002). *Politics and the American press: The rise of objectivity, 1865–1920*. New York: Cambridge University Press.

Manjoo, F. (2017, May 10). Tech's Frightful Five: They've got us. *The New York Times*. Retrieved from https://www.nytimes.com/2017/05/10/technology/techs-frightful-five-theyve-got-us.html

McGlynn, K. (2013). 17 reasons why the kids don't like Facebook anymore. *Huffington Post*. Retrieved from: https://www.huffpost.com/entry/17-reasons-why-the-kids-dont-like-facebook-anymore_n_3825165

Net US digital ad revenues, by company, 2016–2020. *Insider intelligence, eMarketer*. Retrieved from https://www.emarketer.com/chart/222598/net-us-digital-ad-revenues-by-company-2016-2020-billions

Nicas, J. (2022, January 3). Apple becomes first company to hit $3 trillion market value. *The New York Times*. Retrieved from https://www.nytimes.com/2022/01/03/technology/apple-3-trillion-market-value.html

O'Brien, M. (2015, August 11). The real reason that Google changed its name. *The Washington Post*. Retrieved from https://www.washingtonpost.com/news/wonk/wp/2015/08/11/the-real-reason-that-google-is-changing-its-name/

Page, L. (2015, August). *G is for Google. Alphabet*. Retrieved from https://abc.xyz/.

Roose, K. (2021, October 29). The metaverse is Mark Zuckerberg's escape hatch. *The New York Times*. https://www.nytimes.com/2021/10/29/technology/meta-facebook-zuckerberg.html

Smith, D. (2016, December 6). One year later, nobody knows what Alphabet is—and that's a godsend for Google's public image problems. *Insider*. Retrieved from https://www.businessinsider.com/google-vs-alphabet-name-change-public-image-2016-12

Zuckerberg, M. (2021, October 28). Founder's letter. *Meta*. Retrieved from https://about.fb.com/news/2021/10/founders-letter/

CHAPTER 4

The Medium Is the Mass-Age: Revisiting Marshall McLuhan

Figure 4.1. Studies have shown that the mere presence of your phone can be a distraction. Pearls Before Swine © 2020 Stephan Pastis. Reprinted by permission of Andrews McMeel Syndication. All rights reserved.

Takeaway: Turn on, tune in, drop out. The very presence of information communication technologies affects the way we interact with each other.

There are two things to consider with any technology: Invention and mass adoption. Invention is the basic idea of something new being created, tested, and introduced to the public. **Mass adoption** is the idea that an invention only changes culture if a critical mass of people gains access and finds it worthy of integration into their lives.

Example #1: Apple's iPhone was released in June 2007. Four months later it was named *Time* magazine's "Invention of the Year."

Example #2: Google previewed Google Glass in 2012, making *Time*'s list of "Best Inventions of the Year." A year later, *Forbes* magazine predicted that 2014 would be "the year of wearable technology."

On one hand, mass adoption of the iPhone and its smartphone competitors has changed the way the world communicates, with more smartphones than humans on the planet today.

On the other, Google Glass disappeared from the retail market in 2014 after it failed to appeal to U.S. consumers. In fact, wearable technology in general—from digital eyewear to smart watches—has not been enthusiastically received.

When compared to other digital device adoption, wearables have notably lagged in consumer adoption. Some 97% of Americans are estimated to have cellphones, with the majority being smartphones. And 77% of those with mobile phones also have either a laptop or desktop. Only about 21% of Americans own smart watches,[1] a category that also includes fitness trackers (like the Alphabet-owned Fitbit). While smartphone ownership is high across a range of income levels, wearables are far less likely to be used by those with lower education and income.[2]

One term to describe this is **diffusion of innovations**, which is an approach that's been around for about 60 years.[3] The idea is that adoption of technologies can be uneven—some groups are innovators, some are early adopters, early and late majority groups, and finally laggards, who may never adopt.

The simple *existence* of a technology, then, does not have the ability to significantly change culture; mass adoption and intensive use do.

The first air conditioner was invented in the very early 1900s, but it wasn't until after World War II that air conditioners found their way into millions of American homes. Acceptance of air conditioning by consumers—mass adoption—resulted in population booms in the nation's hot spots, from Southern California to Florida. Front porches and sleeping porches as a place to ride out warm evenings were over.

Television sets were available on the commercial market in the United States and elsewhere by the late 1920s, but there was a significant shortage of content. If there's nothing on, why buy a TV? And if few people own TVs, why increase broadcasts? The mass adoption of TV came decades later, also in the prosperous period following World War II.

It's not likely that a family would want to hang out watching Walter Cronkite or Lawrence Welk in a sweltering living room. But add TV to a living room cooled with air conditioning, and a summer haven resulted. In 1950, fewer than 10% of American homes had television; a decade later, nearly 9 out of 10 homes had at least one TV set.[4]

---

1   "About one-in-five Americans use a smart watch or fitness tracker," by Emily Vogels, Pew Research Center (2020).
2   Mobile fact sheet, Pew Research Center (2021).
3   K. Jensen (Ed.) (2012). *A handbook of media and communication research* (2nd ed.), p. 157.
4   Dying to know more about TV history? Wrong book. This about covers our exploration of TV. Tom Genova's TV History is a fabulous resource for info about the first 75 years of TV in the United States: http://www.tvhistory.tv.

Remember This Name: Marshall McLuhan

It was in that postwar TV boom that Canadian media theorist Marshall McLuhan rose to prominence. He was very concerned with the effects of the technology itself, that its very presence would have an impact on people's interactions with each other.

McLuhan coined the term "the medium is the message," which means that the media conveyance device—the technology that delivers the message—would be the agent of social change, rather than the messages that flow through it. His point was that by accepting and adopting a particular information interface, that is, communication technology, we effectively change our communication scale, pace, and patterns.

Prior to air conditioning and television, people had living rooms where furniture was likely to be oriented for face-to-face conversation; after the arrival of TV, furniture tended to be oriented *toward the device.* The television didn't even have to be turned on—by virtue of the technology being in the room, people turned toward the blank screen and away from each other.

By 1970, McLuhan's puns, halting speaking cadence, out-there theories, and *Playboy* magazine interview had cast him as an overexposed media kook. He even had become a punchline on the popular comedy show "Laugh-In."[5] It was a bit ironic that millions of people were watching McLuhan's ideas about TV be mocked on TV.

By the 1980s, his work was mentioned in media courses as mostly a cautionary tale regarding ideas that fail to play by the staid rules of academia. In part, this was because McLuhan's conceptualizations were so esoteric and his writing style so top-of-mind.

So why are you reading about him here? Because today McLuhan is acknowledged as one of the first media theorists to predict the impacts of a global internet, World Wide Web, and smartphones—none of which existed for consumer use by the time of his death in 1980. McLuhan biographer Philip Marchand, who rode the wave of McLuhan's initial rise and fall, wasn't surprised by his return to relevance:

> His insights about the effect of electronic technology in particular—the re-tribalization of the young, the vanishing of such concepts as privacy, the weakening of personal identity, the tendency among users of the media to become what McLuhan called "discarnate," or almost literally bodiless—these insights are more pertinent than ever in the world of Facebook and iPhones. His writings from the sixties and seventies seem to apply more to our own era than they do to his.[6]

Marchand thought the public had largely missed the point: McLuhan wasn't focused on television, *per se,* but on "what this technology was actually doing to our minds and our sensibilities." McLuhan's concern was that reliance on "electric technology" would dramatically change *everything*—family, education, community, work, government—and force us "to reconsider and re-evaluate practically every thought, every action, and every institution formerly taken for granted."[7]

Although television's influence on American culture is irrefutable for many of the reasons cited by McLuhan, his laser-beam focus on television is part of why he became a punchline on a comedy show. As he once said, if you want to save civilization as we know it, "you'd better get an ax and smash all the

---

5   You've never heard of "Laugh-in"? It was kind of the early-1970s version of "Saturday Night Live," featuring sketches and comedy by a variety of popular actors. Ditzy blonde Goldie Hawn had the catchphrase "Marshall McLuhan, whatcha' doin'?" Don't know who Goldie Hawn is? Partnered with Kurt Russell? Still nothing? Mother of Kate Hudson? *Still nothing?* Never mind.

6   You can read Marchand's article in *The Toronto Star* online. Search for "The Fall and Rise of Marshall McLuhan." And, through the time-shifting magic of the World Wide Web, you can even see McLuhan, who died in 1980, explore some of his concepts *in black and white* (search: McLuhan video).

7   McLuhan and Fiore (1967), p. 8.

sets."[8] This was not a popular stance in the 1960s United States, where television was gaining ground at an incredible pace.

McLuhan argued that "television demands participation and involvement in depth of the whole being. It will not work as a background. It engages you."[9]

While McLuhan was correct about the social influence of television as a technology, he was wrong about the level of actual engagement it elicits. Television was and remains a largely passive medium. Short of pausing and skipping, it still is not possible to *interact* with a television program in real time.

Interactivity, in McLuhan's day, was simply too technologically unworkable. Feedback came through ratings, a slow and awkward way to "talk back" to the medium.

Now, of course, more kinds of interactivity are possible, but television shows largely have failed to incorporate that potential.

Look at the difference between *The Walking Dead*—a traditional closed, fixed-narrative program with occasional promotional interaction points—and *Talking Dead*—a fan-group hangout about the show that conducted live polls, had both studio and call-in participation, and, most importantly, acknowledged and discussed what was happening with the audience when they watched the show.

Or consider the sizable and growing number of podcasts that exist as places for people to sit and talk about TV program episodes they have just watched, from "*The Bachelor* Happy Hour" to "The *Succession* Podcast" to "Binge Mode: *Game of Thrones*" and nine other *Game of Thrones* Podcasts. Some of these are starting to be produced in conjunction with the producers or companies making the shows themselves, and this is branching into other media as well.

The main interaction, however, is with the social channel via a computer or mobile device, not the program itself. Watch, don't watch, vote, don't vote, the program goes on with or without you.[10] In contrast, if no one posted on a crowdsourced social media site affiliated with *Talking Dead*, like Facebook or Twitter, the feed would simply stop.

Interaction Makes the Medium Matter More

Perhaps a more appropriate evaluation of TV's influence came from McLuhan with the phrase "turn on, tune in, and drop out."[11] It's likely he was pointing to television's ability to change the context in which we communicate and distract us from ourselves.

He was observing television's ability to numb our brains—turning us away from each other and toward the TV set as our bodies nestle deeper into a comfy couch. Yet there is no participation, just passive acceptance of whatever entertainment offering we choose or happens to be flowing forth from the device.

McLuhan also predicted that the greater the interaction, the greater the distraction. Adding something like a gaming console, the internet, and/or the World Wide Web gives an internet-connected TV—and nearly every other digital device—extraordinary new power to change the way people communicate with one another.

The family home was where McLuhan started his consideration of the impact of technology, and it's still a good place to start in our contemporary exploration of the impact of a given medium.

---

8   Gordon (1997), p. 301.
9   McLuhan & Fiore, op. cit., p. 125.
10 Such efforts at "interactive" TV are not new. A notable example was *Winky Dink and You*, a children's program that aired in the 1950s. The program encouraged parents to buy the interactivity kit, which allowed children to put a plastic sheet over the TV and interact with the program by drawing on the screen when instructed to do so. Draw the right thing, the wrong thing, or don't draw at all—the children's "interactions" were irrelevant to the program's outcome. There's an unintentionally hilarious entry in Wikipedia about why this program was canceled.
11 This was attributed to McLuhan by Timothy Leary and also became a catchphrase for 1960s drug culture. Go figure.

In 2009, a video entitled *Project Natal* was uploaded to YouTube by a marketing team from Microsoft. The video previewed the then-yet-to-be-released Xbox 360 Kinect.

It starts in an empty family living room, where, after a few seconds, the TV turns on, seemingly by itself. A teenage boy approaches the large television set and walks by the TV, where someone who appears to be an animated kung fu master awaits. Sensing the teen's presence, "he" wakes up, follows the teen as he moves around the room, and challenges him to a battle. The teen launches into fight moves, which the game senses.

Next the family of four[12] is shown sitting on the couch side-by-side, playing a driving game. They interact with the game but not each other.

In another scene, the teenage daughter "shops" with a friend for a dress for a dance. The friend appears on the screen only.

Later, the son returns with his skateboard and scans it into the system so he can use "his" board in the virtual game. The board itself is set off to the side, which begs the question, "If you have a skateboard, why don't you go outside and ride it?" The video clearly demonstrates the answer: "Why would you, when Kinect can do it all?"

As the advertising motto for the gaming console proclaims, "The only experience you need is life experience." You don't need other humans, you don't need to go outside, and all communication is best when it's mediated.

The primary purpose of the video is obvious: to promote Kinect. But a McLuhanesque view would see the underpinning message about the medium: It shows the potential Kinect user how to accommodate the technology by organizing the space around it.

The TV needs to be at roughly eye level and the couch at least 10 feet away, which is the distance required for the Kinect sensor.

There are bookshelves, but few books. The kitchen is placed at an angle so the cook can see the TV without stopping meal preparation. Even the refrigerator faces the television. There's a dining room table, but no one sits at it. If the family sits, it's on the couch. There's a bike, but it never moves from its spot beside the dining room table.

The space is dominated by the television set, the invisible gaming console, and, ultimately, the games.

A key thing that McLuhan understood was that every successful technological innovation brings with it a "hidden environment of services,"[13] which in turn has the capacity to change people's behavior. So, it wasn't the TV in the living room that began the electronic revolution, it was the industry of content and connectivity that grew up around it and cemented the set's place in our lives. And it wasn't the iPhone that changed us, it was the integration of the multimodal apps the device ushered into every aspect of our lives.

Hot, Cool, Warm, Whatever: The Interface Controls the Interaction

A communication medium is something that conveys information to us and, while the technology may alter the context in which messages are received, it isn't the *messages* that change but our *responses* to them.

---

12   The demographics of the primary family engaging with the device are notable: white, physically fit, attractive, comfortably prosperous, mom-dad-son-daughter. They play a game with a black family of similar configuration (minus the son). These families signal who the primary market for the device is desired to be—not just *gamers* but families who engage with Kinect, together and apart.

13   M. McLuhan (1970), "Living in an acoustic world," public speech.

The entire notion of "media" is that messages don't come directly to us, brain-to-brain, but are delivered by some *thing* in between. That mediation, that *medium,* as we call it, could be any range of technologies, from paper to LCD touch screens.

McLuhan made a complicated and somewhat confounding distinction between what he character-ized as **"hot" or "cool" media**.[14] This is a concept that adapts well to the 21st century—if you deviate somewhat from McLuhan's relatively fluid (and sometimes contradictory) thinking.[15]

McLuhan was interested in the dynamic interaction between social environment, media user, and medium. It really depended on *how* something was being used, by whom, and in what cultural context. To that end, the electronic transmission of information was less about the meaning of the message and more about the effect the medium would have on the interpretation of the message.

A "hot" medium was one McLuhan termed "high definition," in that it offered an abundance of information, while he classified a "cool" medium as "low definition," in that it required high participa-tion of the brain to reconstitute the information.[16]

A contemporary example of a hot medium is the annual summer blockbuster films that assault the eyes and ears with action and ask the viewer for nothing other than full sensory attention. Engaging the brain typically ruins a good action film, as the majority of the feats are not possible and the plots implausible. Sitting in a theater is a passive, shared experience. You're there with dozens of other peo-ple, but you're not supposed to talk to them.

Photographs also would be classified a hot medium, in that they don't ask you to imagine what's there. Contemporary technology, particularly the ubiquitous use of digital image filters, may distort our sense of what's real, but we don't have to imagine what's there.[17]

Books may well be the quintessential definition of cool media. We physically have to interact with the book to turn the pages and full brain focus is required to follow novel-length plotlines. If you've ever read a book then seen the film adaptation, this likely makes sense to you. We often construct worlds and characters in our minds that don't match up with the choice of an actor or artistic vision of a filmmaker.

While the idea of what constitutes a book today may conjure print, digital, or audio formats, it's worth noting that paper also is an information interface, one that has proven particularly durable over time. There's something about the tangibility of paper versus the intangibility of digitization. Print, by its very nature, is a fixed medium; anything digital can be quickly changed.

Recent studies have shown that we don't retain or comprehend information as well on a digital device as we do on paper—reading online in any form requires a different sort of brain training.[18]

---

14   McLuhan (1964).

15   In short, his perspectives evolved, along with his ideas of where the technology was headed. Here we have simplified and clarified McLuhan's hot and cool concept as it fits with contemporary media. For example, McLuhan initially catego-rized television as *cool* media. It seems, however, a better fit with hot media due to the lack of interaction in interpreting the content.

16   We prefer the term "reconstitute" to "reconstruct" here because the information preexists in some form before we con-sume it, kind of like a can of frozen, reconstituted orange juice. It takes thawing and adding water to actually make the frozen lump orange juice again. The orange juice existed in a prior form, but we determine what it *is to us* through the process of consuming it.

17   It's worth noting here that it's not unusual for the disputed reality of a static photograph to go viral online. One of the best examples is the infamous dress that appeared to some people as blue and black, but to others as white and gold (put white gold dress into a search engine to judge for yourself). These types of optical illusions have been popular as long as there have been eyes to see the images they come from.

18   Amazon's Kindle came out in 2007, just five months after the iPhone. It's striking that it only took seven years of elec-tronic reading to wipe out hundreds of years of human reading habits. *The New Yorker* had an excellent piece on this in July 2014, "Being a Better Online Reader," by Maria Konnikova. It was very hard to read the 2,343-word article without skipping and skimming. Reading online requires different approaches to reading *and* writing

Paper is stable, as it can be read and reread, with nothing changing except your interpretation. As media critic William Powers noted, it's "just this one thing":

> The book you place on your nightstand as you drift off to sleep will be exactly the same book when you wake up in the morning. The newspaper you hold in your hands cannot make an erroneous story disappear as if it had never existed.[19]

Both telephones and comic strips also are excellent examples of cool media, in that you have to fill in the sensory detail that's missing. With the former, you must open space in your mind for engagement that happens through and between the ears; with the latter, you fill in the action between the panels. If you're in the habit of building stories in your mind when you listen to music, then music is a nice fit as a cool medium, as well.

In McLuhan's world, declaring media interfaces as "hot" or "cool" was somewhat easier—a TV was for watching, a radio for listening, a telephone for talking, a book for reading and *always* made of paper.

Once electronic devices began to emerge and their utility started to overlap, all clear hot/cool categorizations began to blur. Digital media—web, mobile, whatever—are like putting all other media in a food processor and blending them into nothing you can easily categorize. Are you still reading a book if you're doing it on a tablet? If you're streaming a movie through your Xbox, is it still a gaming console? Is playing Candy Crush on a smartphone gaming?

The multilayering of device utility has resulted in a rise of what can only be classified as "warm" technology. Multimodal digital devices (like smartphones) are capable of behaving as either hot or cool, depending on how they are being used at any given moment. If you are using it to post to social media or talk on the telephone, a smartphone is cool because it requires your brainpower more than your senses. If you are using it to consume an endless stream of TikTok videos or scroll through a feed of pictures, it's hot.

And, whereas watching a movie in a movie theater or a TV show with a group of family or friends on a shared screen may be considered social, the same isn't true if you're consuming that content alone via smartphone or laptop. In either case you may be absorbed enough not to talk during the show, but with the former you're watching the same thing and can talk about it later. The very nature of today's digital technology is that it isolates us within our needs, wants, and interests, including multitasking with multiple devices, even when we're with others.

McLuhan said people are in a state of perpetual modification as technology extends parts of the physiological self. And each technological innovation nurtures those that inevitably come after it, much as a pollen-spreading bee not only perpetuates a plant species but also has the capacity to expand and improve it.[20]

In response, "the machine world reciprocates man's love by expediting his wishes and desires, namely, in providing him with wealth."[21] In contemporary digital media terms, that would be best rephrased as *"a wealth of information."*

And as we embrace our technologies, we also must serve and maintain them. McLuhan gave the analogy of a cowboy and horse or the executive and clock—the tools become the taskmaster.

---

19   Powers (2006).
20   In case this is confusing: People are the bees; machines are the plants. As McLuhan (1964) put it, "Man becomes, as it were, the sex organs of the machine world, as the bee of the plant world, enabling it to fecundate and to evolve ever new forms" (p. 46). That guy really could turn a phrase.
21   McLuhan (1964), op. cit., p. 46.

A more contemporary analogy would be the individual and social media feeds or personal email. While consuming either may seem like a harmless pastime, it also can be incredibly anxiety inducing if you are delayed in your engagement. There are two reasons for this anxiety, which work in tandem:

- Fear of falling behind on the ever-accumulating content of feeds and email; and,

- Fear of missing out, FOMO.

Taking vacation off the grid (e.g., backpacking in the wilderness) can result in the seemingly endless time suck of catching up.

How many emails do you have in your inbox? A few? Hundreds? More than a thousand? Why? Because email is like triage, we focus on what *has* to be addressed first and leave the rest for that never-happening time of later.

It's important to note that the challenges of and adaptation to technological change are not experienced equally by all. If you don't perceive a need to learn or adopt a particular technology, or have the desire to do so, you may choose not to. This is a luxury generally afforded only those whose education and jobs were solidified prior to technological change—and people who have the time, perhaps the greatest commodity we forget to discuss!

Someone in their 50s may never learn the intricacies of digital media management and will see little to no job impact, while someone in their 20s may be expected to lead on it as soon as they're hired. McLuhan pointed to this liminal place between technologies as a difficult place to be—you must decide quickly when to adopt and how to adapt. As he said,

> There are very many reasons why most people prefer to live in the age just behind them. It's safer. To live right on the shooting line, right on the frontier of change, is terrifying.[22]

True? Not True? Does It Matter?

Legend has it that Albert Einstein[23] would spend days focused on portions of mathematical formulas. His depth of attention to and interest in the subject at hand led him to revolutionary thinking, which was facilitated by his ability to get into a state of *flow.*

Flow is the act of getting completely immersed in something with no distraction. It's often attributed as generating true creativity. In fact, people who do research often report epiphanies that wake them in the middle of the night. It's that experience of "aha, this is how it fits together" that results from your brain continuing to work through a problem even after you've consciously detached yourself from it.

This is why it's a good idea to have an audio recorder or notebook and pen at bedside when you're working on a project or trying to solve a problem. If you wait until morning, your mind will have moved on and the solution will be about as clear as the story line of a dream.

If Einstein were working today, would he get sucked out of flow by a warm technology world of email, Web surfing, YouTube videos, and/or tweets? Would text messages at all hours prevent him from achieving deep sleep?

Quick show of hands: How many of you sleep with your smartphone at your bedside? Of course you do! It's also your alarm clock. And how many of you wake and check text messages and social

---

22   Ed Fitzgerald interview with Marshall McLuhan on the CBC's "New Majority." (1970)
23   If your first reaction was "Who?"—please do a search and find out who this is. He's important.

media in the middle of the night? Of course you do! Disrupting the potential for epiphanies is a solid reason why you should not.

Digital technology is full of potential for distraction. The brain is repeatedly rewarded for having a short attention span. Devices flash. They chime. They demand attention.

At best, we skim when we attempt to read online. At worst, we disappear into something else, something more entertaining, and the purpose for starting to read in the first place is long forgotten. Digital media, which tend to reward the brain for having a short attention span, have great potential for stifling deep thought. *Squirrel!*

McLuhan thought technological evolution would take us from a **visual world** to what he termed an **acoustic world**.

The visual world held sway beginning in the centuries after the introduction of the Gutenberg press, which ultimately brought print—and literacy—to the masses. The visual world was led by experts, anchored in history, and propagated generally agreed upon truths. Books, then, would be the exemplar of the visual world. As McLuhan noted, "things stayed put. If you had a point of view, that stayed put."[24]

The acoustic world is the result of technologies that bring us everything at once, so even when we're seeking specific information we're surrounded by distracting noise. It is a world characterized by continual change, and everything, including truth, is a moving target and subject to reinterpretation in the absence of context. McLuhan pointed to the acoustic world as a cacophony of information with "no continuity, no homogeneity, and no stasis. Everything is changing."

The visual world was characterized by objectivity, while the acoustic world is one of subjectivity— "immersion without any point of view." And that, McLuhan, said, would change people's orientation from pursuit of concrete goals to individualized distraction: "The acoustic man never has a goal; he just wants to do his thing, wherever he is."[25]

There's a quote attributed to Einstein that is a dead-on-perfect fit for our present fixation with digital devices: "I fear the day that technology will surpass our human interaction. The world will have a generation of idiots."

Part of the reason it's so perfect is that there's *no evidence that Einstein said this.* It started making the rounds on the Web in about 2012, superimposed over a portrait of Einstein. That makes it either a mistake or a **meme.**[26] Either way, it's a great example of the ability for anyone to post anything on the web and for others to take it as True.[27]

Applying McLuhan to Our Technology Usage Patterns

McLuhan's argument was that we like to see ourselves reflected back at us, which draws us to others most like us—an **endless re-presentation of self.** Social media was not created for this purpose, but has evolved to be a great amplifier of individual worldview. We don't necessarily recognize that the online world we create via social media channels is nothing but me, me, me. My preferences. My privacy settings. My politics. My work. My hobbies. My friends.

---

24 McLuhan (1970), op. cit.
25 Ibid.
26 A meme is a video clip, image, hoax, joke, whatever, that goes viral—spreading from person to person—for sometimes inexplicable reasons. One of the most enduring is "Rick Rolling." Look it up, but don't pass it on. It's been around so long that odds are whoever receives it will think you're a dork.
27 Before you send, circulate, or share something interesting from the web, carefully consider the source, then check it out for yourself by consulting multiple sources. Snopes.com exists just for this purpose. If it's an image, take advantage of a reverse image search to see where else the image has been posted. These are the foundational basics of media literacy, which is explored in Chapter 6.

Social media do not require you to make overarching choices about these things from the outset; they allow your feed to evolve via the people, pages, and other feeds you choose to follow. It takes a lot of little choices to make up a Facebook, Twitter, TikTok, or Instagram feed—and, like snowflakes, no two are alike.

Even if you understand that this is what's happening, which you likely do, would you want it any other way? Do you want to see content that offends your closely-held beliefs and challenges your personal perspectives? What do you do with that content? Consume it and broaden your worldview, or use platform tools to hide, block, and mute it?

McLuhan may not have seen social media coming, but he understood that changes in technology would result in our private and social lives—things previously held close to our physical selves—being "hoicked up into full view."[28]

We may spend so much time trying to keep up with everyone else's postings from their own little technology-anchored worlds that we are numbed to things and people existing beyond. Again, McLuhan said this was predictable. As the cycles between information communication technology invention, availability, and innovation compress, we are caught in a seemingly endless cycle of *adopt, engage, adapt, repeat.*

McLuhan pointed to the rising reality of "**womb-to-tomb surveillance**," which would cause "a very serious dilemma between our claim to privacy and the community's need to know." So much personal sharing via electronic media would result in a digitized chronicle of our lives akin to "one big gossip column that is unforgiving, unforgetful and from which there is no redemption, no erasure of early 'mistakes.'"[29]

It's not hard to apply that perspective to contemporary digital media. Consider your own digital identity: Where does your digital footprint begin? It's possible your first digital presence was an ultrasound image shared on social media. From there, close family and distant relatives alike may have posted pictures of you long before you had the ability to consider whether you wanted those pieces of your life online in perpetuity.

And, today, friends and strangers alike can spread images and videos of you without your consent or even knowledge, facilitated by tech that allows near-instant sharing. Think of anyone whose image has been co-opted for a popular meme, or whose former significant other posted intimate images online. How hard is it to make that go away? A digital dossier might be something we could edit and control; contemporary digital media is closer to a digital tattoo.

The Human Quest for Tribes

People have joined together for thousands of years based on blood, location, or common interests. This is known as **tribalization**, and it's been one of the greatest predictors of human success—and failure. Unification into tribes of people who support and protect each other is at the heart of innovation and civilization. On the flip side, tribalization also can pit tribes against one another, resulting in strife, unrest, and war.

A key point in the plotline of the AMC+ original series "Moonhaven" (2022), was that the Earth had all but been destroyed by tribalism, which sows division. As such, members of the Moon colony created to save humanity were forbidden from forming permanent family units based on blood. Monogamy was discouraged and children were given to strangers to be raised as their own.

---

28 Isn't "hoicked" a great word? It basically means being yanked, or pulled in abruptly. McLuhan, op. cit., p. 47.
29 This is a particularly prescient passage from *The Medium Is the Massage* (p. 12). In 1967 McLuhan only saw the potential for such abuses; 50 years later, a quick search of a person's name by a potential employer may reveal youthful indiscretions that can cost that person an adult job.

Interestingly, these rules were made by IO, an artificial intelligence entity that governs virtually all of the settlers' enculturation. Despite direction by a non-human entity professing the best intentions for all, the Moon residents fracture into tribes and violence ensues. The message seems to be that once division is sown, people are incapable of uniting across it. And that's also a classic plotline running through the human opus of creative works.

Think of Shakespeare's famous tragedy "Romeo & Juliet": The Montagues and the Capulets are mortal enemies, locked in an age-old feud. That's a basic form of tribalism—blood allegiance—with tragic consequences.

But let's get back to McLuhan, and how he thought technology would affect tribalization.

He theorized that the world would figuratively shrink as technology allowed people to communicate across time and space, forming "the global village." We would effectively be de-tribalized, as common perspectives and a deeper understanding of our world took root. McLuhan, viewing this change through the lens of the rapidly expanding influence of television, noted that "our new environment compels commitment and participation. We have become irrevocably involved with, and responsible for, each other."[30]

That would not last, however. McLuhan predicted that the same technologies that allowed us to share our realities across the globe, would also overwhelm us with a "worldpool of information."[31] The result would be the eventual fragmentation of common culture, an identity quest he equated to **retribalization.** Presented with an endless flow of information—a veritable tsunami of information—we ultimately would retreat to higher ground. In short, we'd feel compelled to reclaim a tribal existence.

Today, information overload, coupled with an endless desire for entertainment, actively drives us toward our own feeds, where we comfortably consume ourselves and the perspectives of people like us. We repeatedly turn to and are driven toward information that not only fails to challenge our minds but also reflects our own deep-seated beliefs right back at us—*an endless re-presentation of self.*

We don't have to find our tribes; the technology does it for us. And, while the creators of algorithm-driven technologies like "social" media seem shocked at the outcome, their inventions have forged deep cultural divisions. Tribalism is rampant, understanding and empathy are in short supply, and we're well on our way to the technology-driven period of anarchy and cultural collapse that McLuhan predicted nearly six decades ago.

McLuhan's prescient understanding of the collision between human behavior and rapid technological change are why he gets his own chapter in this book.

Summary

Revisiting the work of media critic Marshall McLuhan, who made his mark in the 1960s, puts technological change into perspective as an ongoing, organic, social process. McLuhan argued that the medium—the devices through which we access information—has a larger social impact than the message—the content we receive via a particular medium. Part of this is the ability of technologies, from television programs to social media apps, to reflect back our own perspectives and worldview, rather than challenging either. McLuhan predicted that advancements in media technologies would lead to increased monitoring of our actions and activities, as well as greater interaction with those who share our opinions and beliefs.

30 McLuhan and Fiore (1967), op. cit., p. 24.
31 McLuhan and Fiore (1967), op. cit., p. 14.

Key concepts

Acoustic world
Diffusion of innovations
Endless re-presentation of self
Flow
Hot versus cool media
Mass adoption
Meme
Retribalization
Tribalization
Visual world
Womb-to-tomb surveillance

References

Fitzgerald, E. (1970). Marshall McLuhan interview. *Canadian Broadcasting Network*. Retrieved from https://www.cbc.ca/pla
yer/play/1081198147655

Gordon, W. T. (1997). *Marshall McLuhan: Escape into understanding*. Toronto: Stoddart.

Jensen, K. (Ed.) (2012). *A handbook of media and communication research* (2nd ed.). New York, NY: Routledge.

Konnikova, M. (2014, July 16). Being a better online reader. *The New Yorker*. Retrieved from http://www.newyorker.com/scie
nce/maria-konnikova/being-a-better-online-reader

Marchand, P. (2011 July 19). The fall and rise of Marshall McLuhan. *The Toronto Star*. Retrieved from http://www.thestar.
com/opinion/editorialopinion/2011/07/19/the_fall_and_rise_of_marshall_mcluhan.html

McLuhan, M. (1964). *Understanding media: The extensions of man*. New York, NY: McGraw-Hill.

McLuhan, M. (1970). Living in an acoustic world. *Public lecture, University of South Florida*. Retrieved from https://marsha
llmcluhanspeaks.com/media/mcluhan_pdf_6_JUkCEo0.pdf

McLuhan, M. & Fiore, Q. (1967). *The medium is the massage: An inventory of effects*. New York, NY: Bantam Books.

Mobile fact sheet. (2021, April 7). *Pew Research Center*. Retrieved from https://www.pewresearch.org/internet/fact-sheet/
mobile/

Playboy interview: Marshall McLuhan. (1969, March). *Playboy Magazine*. Retrieved from http://www.nomads.usp.br/leuph
ana/mcluhan_the_playboy_interview.pdf

Powers, W. (2006). *Hamlet's Blackberry: Why paper is eternal*. Joan Shorenstein Center on the Press, Politics and Public Policy,
Discussion Paper Series, Harvard College. Retrieved from http://shorensteincenter.org/wp-content/uploads/2012/03/
d39_powers.pdf

Project natal. (2009, June 1). *Uploaded by XboxE3*. [Video]. Retrieved from https://youtu.be/g_txF7iETX0

Vogels, E. A. (2020). About one-in-five Americans use a smart watch or fitness tracker. *Pew Research Center*. Retrieved from
https://www.pewresearch.org/fact-tank/2020/01/09/about-one-in-five-americans-use-a-smart-watch-or-fitness-tracker/

CHAPTER 5

We're Not *Here:* The Cultural Consequences of All Me, All the Time

Figure 5.1. Studies have shown that the tools our phones provide effectively replace on-brain retention of things like telephone numbers and upcoming appointments. Frequent use of your phone's GPS may also contribute to the shrinking of your hippocampus—the part of your brain that helps you navigate the world. Pearls Before Swine © 2019 Stephan Pastis. Reprinted by permission of Andrews McMeel Syndication. All rights reserved.

Takeaway: Being fed a steady diet of "me" is a key characteristic of the post-information age. But that doesn't make it a good thing.

Does the term **"Pavlov's dog"** ring a bell?

For those who may have missed the (admittedly) bad pun in the preceding sentence, let's quickly review this 100-plus-year-old contribution to behavioral science. It's entirely relevant to this chapter.

Ivan Pavlov was a researcher who studied conditioned response using dogs. One of his lab techs fed meat powder to dogs so they could study the animals' digestive systems. Pavlov discovered after a short time that the dogs would start drooling when they saw the lab tech walk through the door—before the food even came out.

Pavlov focused on that drooling and how the dogs' responses were conditioned by an action (the lab tech's arrival and anticipation of food). Pavlov then tried ringing a bell before he gave a dog the meat powder. After a few tries, the dog began to drool when the bell rang, purely in anticipation of the meat powder.

Pavlov called the drooling-for-the-bell phenomenon a **"psychic secretion"**—basically a learned, unconscious response to a stimulus.[1] It comes down to this response from Pavlov's dog: You show me food, I drool. You ring a bell and show me food, I drool. You ring the bell, I drool.

Surely you can see what's coming here … each digital device and service platform has its own trademark pop, ping, chime, trill of notification—or can be assigned one. We quickly learn what sound goes with what service platform and what device. The iPhone Marimba ring is a great example: You hear that trademark ring and may check your pocket and your iPhone to see if it's *you* that's ringing.

The bell rings and the psychic secretion begins. It may not be drooling, but it's hard to argue that there's no central nervous system response—especially if you can't get that hit of cyberinteraction your mind is expecting.

How long can you go without checking your smartphone? How do you feel when you leave it at home or misplace it? Do you have sweaty palms? An increased heart rate? Anxiety? The stimulus of instant gratification can be quite addictive—and not having it can feel much like drug withdrawal.

Smartphones are all but extensions of some people's hands, hearts, and brains. In fact, we carry so much around in these tiny computers that the U.S. Supreme Court ruled in 2014 that law enforcement could not search them without a warrant. The court, peopled by justices not generally known for their technological savvy, noted the following:

> The term "cell phone" is itself misleading shorthand; many of these devices are in fact minicomputers that also happen to have the capacity to be used as a telephone. They could just as easily be called cameras, video players, rolodexes, calendars, tape recorders, libraries, diaries, albums, televisions, maps, or newspapers.

> One of the most notable distinguishing features of modern cell phones is their immense storage capacity. Before cell phones, a search of a person was limited by physical realities and tended as a general matter to constitute only a narrow intrusion on privacy.[2]

In the past we kept most of our lives in stable documents that stayed in our homes or at work. It wouldn't have been conceivable for a person to carry around a calendar, camera, diary, photo album,

---

1   There's a lot more to Ivan Pavlov's work and we don't want to disrespect it by giving it short shrift, but the science is beyond the point. The psychic secretion is the point. And isn't that a spectacular term?

2   U.S. Supreme Court, June 25, 2014, *Riley v. California*, p. 17. Search for it.

address book, album collection, computer, *and* VCR every minute of every day. Now all of those things are instantly accessible through a smartphone, which takes up less than a square foot of pocket space.

In the past, all of those things likely would have been in your home, which law enforcement could not search without a warrant. Now they must have a warrant to search your smartphone.

Between your Pavlovian response to digital chimes and the Supreme Court's reminder of the minute-to-minute record of our lives that smartphones contain, it's no wonder that a waterlogged, broken, misplaced, or stolen smartphone is a modern tragedy of Shakespearean proportions.

How Deep Does This Technological Ocean Go?

Next time you walk through the grocery store, go to a family restaurant, or take a flight, count the number of children between the ages of 2 and 8 who are keeping time with a parent's smartphone or tablet.

But wait, what about all those kiddos age 9 to teens? They're all tapping away on touch screens, too, as many children over the age of 9 have their *own* cell phones and tablets.

This is a generation of children who expect digital engagement. They see their parents check email, text, update social media posts, and sometimes even talk on the phone while they're stopped at a long traffic light, in line just about anywhere, or waiting at the doctor's office. We're a culture that wants to be digitally distracted every spare minute.

Take, for example, actress/mom Rosemarie DeWitt, who cheerfully shared an anecdote about enticing her infant daughter to crawl by putting her iPhone across the room. The result? "She crawled across the room to the phone. It's not because she'd ever held it … She sees it take our undivided attention."[3]

In short, we're enculturating a generation of digital consumers. **Enculturation** is a process in which you learn what a society values and how you are expected to behave. Parental mediation can help or hurt this process, like when parents are co-viewing, or using instructive or restrictive viewing—when there is attention to screen time at all. Much enculturation is absorbed through family interaction, ingested as we consume media, taught in school, and learned from friends.

And the prevalence of digital devices on all fronts is changing virtually every aspect of our communicational lives.

A Recent Past of Significant Cultural Shifts

It's relatively easy to see digital media through the lens of how *you* use them. A key goal of digital literacy, however, is getting you to think more critically about how you use media, as well as how others use digital media, what the recent past held, and what the fast-upon-us future might hold.

The United States has rapidly transitioned from an industrial society to an information-driven one, from need for brawn to demand for brain in less than two generations.[4]

The initial shift began in the 1950s with mass adoption of television and other electronics (including calculators), which marked the arrival of the **electronic age.** This era was characterized by great optimism that televisions, computers, and men on the moon would usher in the future of robots, flying cars, and space travel that the world had been waiting for.

Instead, the electronic age stagnated for more than 25 years, with most significant communication advancements relating to things like color TV broadcasts and cable television.

---

3   "Movie Stars Grapple With Family Life in the Digital Age," by Andrea Mandell, *USA Weekend*, October 1, 2014.
4   A generation typically is about 25 years. One generation begins at birth and follows the generation that conceived it.

By the mid-1980s, the rise of personal computing and aggressive marketing by Online Service Providers had ushered in the **information age,** which is characterized by individual empowerment and explosive growth of shared knowledge via the internet and, later, the World Wide Web.

Notable achievements during the information age, which ran from roughly 1984 to 2004, included online chat rooms and bulletin-board services, Wikipedia, Google as a search engine, Yahoo!, news aggregation, music, game, and movie downloads (Napster, iTunes), online shopping (Amazon, eBay, craigslist), and the introduction of social media (Friendster, Myspace).

The information age morphed into the **post-information age** somewhere between 2004 and 2008. This shift was facilitated by leaps in technological innovation (smartphones, most notably), mass engagement with social media (LinkedIn, Flickr, Facebook, YouTube, Twitter, Tumblr, Pinterest), and streaming video (Hulu, Netflix).

Well before its arrival, computer scientist Nicholas Negroponte[5] outlined **three characteristics of the post-information age**:

1. Digital living, or "place without space";
2. Asynchronous communication; and
3. Time shifting, or "demanding on demand."[6]

Digital living is the idea of communicating from wherever you happen to be: home, commuting, work, weekends, vacation. People can respond to email and update websites and social media from a place and on a schedule that is very much determined by the individual.

Digital living is facilitated in part by **asynchronous communication,** that is, communication that is literally out of sync—not in real time. For example, I get out of class and text you, but you're still in class. You get the message and respond at your convenience. Email functions in much the same way.

Asynchronous communication is not novel; letters, for example, have always been asynchronous. What is unique to the post-information age is the speed at which individual messages travel.

Synchronous communication, by way of comparison, would include face-to-face communication, telephone conversations, and video messaging (using systems such as Zoom, Microsoft Teams, Facebook Messenger, and FaceTime).

A rapid text conversation or instant messaging can feel somewhat like real-time back-and-forth communication, but there's a key difference: You can leave the conversation at any point and it might take the other person a few minutes to realize you are no longer there. If you do the same thing during a face-to-face conversation, for example, it would be very awkward and socially inept.

While asynchronous communication is about the sending and receiving of digital messages, **time shifting** is about the consumption of digital media offerings when and where you want them.

The convenience of time shifting started with the advent of VCRs and has expanded significantly via the introduction of DVRs and video streaming services, like Netflix. If you don't want to stay up until 10 p.m. Tuesday to watch your favorite show, you digitally record it and watch at a time that suits you. And if you want to binge-watch an entire season of *Breaking Bad* in one 24-hour period, you can (whether or not you *should* is another issue).

---

5   From the remarkably forward-thinking book, *Being Digital* (1995), Chapter 13: "The Post-Information Age" (pp. 163–171).
6   Interestingly enough, Comcast/Xfinity's streaming video service is named "On Demand." Coincidence?

Speaking of Time …

You've heard of leap year—a day tacked onto February once every four years—but have you heard of the leap second?

The **leap second** is a complicated way of ensuring that the time that governs every waking minute of our contemporary lives matches the earth's rotation.

We haven't always been directed by exact time keeping.[7] Analog clocks and watches rarely matched each other and required synchronization—putting them on the same time—to come close.

Think of the diminishing range of clocks you have to move forward and back at the beginning and end of daylight saving time. It's hard to get your microwave and oven clocks to consistently display the correct time, because they're manually set, not receiving a digital signal that automatically adjusts the time.

There's actually an organization that maintains global time, the International Earth Rotation and Reference Systems Service (IERS). Some version of that agency has existed since the 1800s.

One role of the agency is to determine and announce the need for a leap second. While this all gets very complicated very quickly, the basic idea is that the Earth's rotation infinitesimally surges and lags, while atomic and digital clock time remains exactly the same.

The solution is to either speed up or stop all clocks that receive a digital signal by one second, as needed. Some 27 single-second adjustments have been made by global timekeepers since 1972, the last of which was December 31, 2016.

The problem: The increasingly sophisticated computers that underpin, well, everything, don't do well with the leap second. In 2012, the addition of a leap second crashed Foursquare, LinkedIn, Reddit, and Yelp.[8]

As a result, there's concern every year as the IERS evaluates data and determines whether to add a leap second June 30 and/or December 31.[9] And there are some folks who are very interested in getting rid of the leap second—notably Alphabet, Amazon, Meta, and Microsoft.

Prior to June 30, 2022, Meta's production engineering team called for use of newer, more accurate technologies to tell time, predicting future internet outages if the leap second continues. They explained,

As an industry, we bump into problems whenever a leap second is introduced. And because it's such a rare event, it devastates the community every time it happens. With a growing demand for clock precision across all industries, the leap second is now causing more damage than good, resulting in disturbances and outages.[10]

Meta's perspective points to potential changes for the future and the fact that our biggest tech companies literally have influence over time.

---

7   You know that old adage, "even a stopped clock is right twice a day"? The idea is that even something that generally provides bad information is occasionally right. But we digress …

8   "Leap seconds cause chaos for computers—Meta wants to get rid of them" by James Vincent, *The Verge*.

9   While the rotation of the Earth has been infinitesimally slowing for decades, there's increasing evidence that the rotation now is speeding up—June 29, 2022, was the shortest day since atomic clock collection began in the late 1960s. As we noted, this gets complicated quickly and we don't want to get too far out in the weeds. If you're interested in learning more, go to your favorite search engine and start searching.

10   "It's time to leave the leap second in the past" by Oleg Obleukhov & Ahmad Byagowi, Meta production engineering.

What Could Be Better Than All Me, All the Time?

Taken together, digital living, asynchronous communication, and time shifting comprise **"The Daily Me."** Negroponte pointed to this as the difference between narrowcasting and digital.

- **Narrowcasting** is part of the information age, as it represents a scaled down version of the mass media model. Narrowcasting shoots for the individual and may or may not hit the intended party—it's the idea of personalization as envisioned by a conglomerate media outlet.
  o You're targeted largely on demographics; the message is not being sent to *you*, it's being sent to a whole bunch of people someone thinks are *like you.*
- **Digital** is a hallmark of the post-information age. Messages aren't being targeted to people *like* you, they're being sent *to you.* It's actual personalization by me and for me; I am in control and I determine what information pathways I follow and when I follow them.
  o Web platform service providers happily play into this by tracking your every move and sending hyper-personalized advertising to follow you around on the Web. Remember those shoes you left in the shopping cart at Zappos? They're still there waiting for you—and showing up on every ad-zap enabled page you visit until you cave in.

Three characteristics of The Daily Me, which you may recall from Chapter 2's discussion of the third wave of digital development, are partnership, participation, and personalization.

Partnership is between the Platform Service Provider and the end user. We agree to a site's terms of use—user agreements—in exchange for free use of a given service. And it's quite common that users agree to these "agreements" without even reading them.

Such approval is the essence of *imprimatur,* a lovely Latin word that translates to "let it be printed." In this case, the "it" is the content we publish on platforms that reflects individual interests.

Via user agreements, we give permission for that content to be mined for personal details. This allows for the hyper-personalization of advertising, based entirely on user engagement across the web. In short, we allow companies to track us, in exchange for free access to services such as social media sites.

Participation in the post-information age is a direct result of partnership. In exchange for demographics and our digital privacy, we are provided a space to participate—to share our observations, photos, videos, music, and a great deal of decidedly unoriginal content. That makes us both content provider and content consumer, with digital service companies acting as host—a partnership that works well as long as we're willing to ignore its social costs.

Personalization, which is enabled only by partnership and participation, takes three forms:

1. *Individuation*
   - We choose what content to add to our pages and feeds, which differentiates our space on platforms from all others. A social media feed, for example, is unique to its creator. We can live in very different virtual worlds from those closest to us in our physical lives.
   - We are offered templates which we can use that have different styles, but come as a set of pre-organized elements. We may not be able to change typeface or formatting, but we can add our own pictures, thoughts, favorites.
   - We build an "us" across our digital media use, made up of all the little pieces we pull into the picture.
2. *Selection*
   - We choose whose content we want to see and who can see ours.

3. *Validation*
    – An endless cycle of picking content that reflects us, getting feedback from those who see it, and having our interests reinforced by seeing algorithm-driven "suggestions" for content similar to our own.

When coupled with mobile devices, the result is a system built for distraction and information overload.

In other words, the benefits to digital living are clear: My workspace and a world of information are with me wherever I am. The drawback is equally clear: I live in a world where nonstop data pervade every aspect of our lives.

Unintended Effects of Mediated Life

When our days are clogged with text messages, social media feeds, and dozens (if not hundreds) of emails, what are we actually accomplishing? Are we building and maintaining real relationships or digital replicas?

In her book, *Life on the Screen*,[11] Sherry Turkle noted that our infatuation with The Daily Me results in the belief that digital interactions are somehow authentic and significant to our overall lives. She pointed to **three deceptive effects of digital living:**

– **The Disneyland Effect**
   One of the primary goals of "The Magic Kingdom" is and always has been to help guests suspend reality for a while. When we're riding on the Jungle Cruise or Haunted Mansion, we're supposed to believe we're really in a jungle or a haunted mansion. It's part of the experience. The Disneyland Effect is *making artificial experiences seem real.*
   One need only think of Facebook connections. They're called "friends," yet interactions typically are very one sided, often borderline bragging, and life's inevitable bad news is shunned (there is no "dislike" button). We collect "friends" and follow their lives in our feeds—often building on a relationship that hasn't existed, doesn't exist, and likely won't ever exist.

– **The Artificial Crocodile Effect**
   Have you ever seen a crocodile purse? Belt? Boots? Crocodile has an interesting texture that is prized by some designers as an exotic skin. But, as the leather ages, it tends to crack and peel. It's extremely expensive and needs special care to last. Why not replace it with artificial crocodile, which is made of durable vinyl and nearly indistinguishable from the real thing? The Artificial Crocodile Effect is *making the fake seem more attractive than the real.*
   Text messaging is awesome. I don't have to actually talk to anyone and can ignore messages from people and on topics I'd like to ignore. By comparison, face-to-face interactions offer nowhere to hide—and might even require eye contact.

– **The Popularity Effect**
   How do we judge people on social media? Number of followers? Shares? Retweets? Friends?

---

11 S. Turkle (1995). *Life on the screen: Identity in the age of the internet.* Turkle is a professor at MIT and founding director of the university's Initiative on Technology and Self. She has written several other books about how the presence of digital devices changes our world, including *Alone Together* (2011) and *Reclaiming Conversation* (2015). She knows this stuff.

The Popularity Effect is *finding virtual life so compelling that we think we are actually accomplishing something.*

But here's the core question: Can you put your number of Facebook friends on a résumé, or is it like your GPA (i.e., no one cares how many Facebook friends you have)? If working with social media is part of your job, the answer is probably "yes." If not, don't even bother trying *because no one in the physical world cares.*

Turkle, who is founding director of MIT's Initiative on Technology and Self, asked why digital living and real life must compete. "Why can't we have both?" she asked, then answered her own question: "The answer is of course that we will have both. The more important question is, 'How can we get the best of both?'"

All Me and the Algorithms—All the Time

Not all of what happens online is determined by a particular person acting synchronously. Much of what happens is the result of **algorithms**, which are computer programs that massively accelerate decision making and associations. They are programmed to sort things out very quickly so that associations can be made.[12]

Algorithms, for example, can look at what music you like, and look at other people who like some of the same things, and make suggestions for other things you might like. They do the same thing with videos, movies, websites—basically any activity where data can be gathered and patterns of associations can be compiled.

It's important to keep in mind that those who program algorithms are human and bring all of their cultural constructions with them when they design the algorithms. Their efforts also frequently are motivated by companies that prioritize innovation and profit over cultural consequences. So if a designer's goal is to get you to keep watching, to stay on a particular platform or media channel, they will harness the power of the algorithm to keep you watching.

If you keep watching, and the algorithm was designed to keep you watching, the rules of success have been met. Whether that's good for you or for society in general is another question.

We can think of this as the ability of algorithms to have an amplifying effect on our experiences. Personalization algorithms work to bring more attention to some voices on social media and reduce attention to others. A study in 2021 found that the personalization algorithms amplify sources that are more partisan—politically further to the left or right—compared to sources more in the center.[13]

This all has to do with how people and computing through media work together. To get a grasp of that it is always a good critical move to see how such decisions were started, who was behind them, and whether the results were what those inventors desired.

Tim Berners Lee and the Semantic Web

Tim Berners-Lee is credited as the inventor of the World Wide Web. Later he proposed the idea of the **Semantic Web**, which seeks to make the meaning of data machine-readable. In 1999, Berners-Lee said he had a dream that all the content, links, and transactions on the web would be machine readable, so that eventually trade, bureaucracy, and our daily lives would happen with machines talking to other machines.

Algorithms, as they grow in sophistication, are approaching such a dream. The meanings and associations between things we do and say online feed into massive data analytics so that our actions

---

12   There is an excellent introduction to algorithms in a BBC documentary called *The Secret Rules of Modern Living: Algorithms*, hosted by mathematics professor Marcus Du Sautoy. It can be found on YouTube.

13   Huszár, Ktena, O'Brien, and Hardt (2021)

become machines talking to other machines, using sets of data based in our actions to anticipate and shape our experiences.

Who Benefits From Algorithms?

We can see the ease with which the algorithms make things easier in many ways, but we can become blind to the way our minds are shaped by these experiences. Because these are not always to our benefit.

So it's important to keep an eye on who really does benefit from these algorithms. If the goal is to keep us watching so that we are exposed to more ads, then whoever is making money from our data and attention is a key beneficiary.

As end users, we are in less of a position to benefit. Maybe we gain pleasure from the ads, maybe we don't feel like our time is being taken from us. But then again, maybe we do; and maybe, in the effort to keep us watching, we are being shown things that are engaging, but not really good for us.

That may be the most surprising turn of All Me, All The Time. As more and more decision making moves to the web of machines that are effectively creating and undermining meaning, the less we are actually in control.

How Does All Me, All the Time Affect Your Health & Well Being?

Beyond apps and algorithms, let's think about the device that holds them all. Smartphones are designed for convenience. They've collapsed dozens of functions that used to take up space in the physical world into one handy, hand-held device. Think telephones, clocks, notepads, calendars, cameras, computers, televisions, banks, stores, and so on.

Smartphone and app developers are highly motivated to keep you skimming, watching, scrolling, ad infinitum. They're attracted to innovation because keeping your attention is extraordinarily profitable. And we can't / won't / don't want to stop. Our attention to devices tends to creep in where ever we are—at dinner with friends, in the classroom, in a Zoom meeting. Marketing professor Adam Alter noted that "life is more convenient than ever, but convenience also has weaponized temptation."[14]

As we noted at the beginning of this chapter, our bodies develop a physical reaction to digital devices. As with other pleasurable stimulants, that physical reaction may ultimately shape itself into **behavioral addiction**, or the unending pursuit of instant gratification.

Alter defined behavioral addiction as "something you enjoy doing in the short term, that undermines your well-being in the long term—but that you do compulsively anyway."[15] And, humans are hard-wired for addiction:

> We are engineered in such a way that as long as an experience hits the right buttons, our brains will release the neurotransmitter dopamine. We'll get a flood of dopamine that makes us feel wonderful in the short term, though in the long term you build a tolerance and want more.[16]

Dopamine is one of three neurotransmitters—catecholamines—produced by our bodies. The other two are adrenaline, which you might recognize as causing a "fight or flight" reaction when triggered, and norepinephrine, which regulates your blood pressure, among other things.

Dopamine taps into our pleasure centers. Technology is designed to tap into our pleasure centers. You get a bit of a dopamine hit from things like social media likes and text notifications—all of which today flow through our smartphones. As Alter noted, behavioral addictions tend to emerge when

---

14   Alter (2017), p. 19.
15   Interview with Adam Alter in "Why we can't look away from our screens" by Claudia Dreifus, *The New York Times*.
16   Ibid.

something we enjoy fulfills a need, you can't imagine being without it, and you use it to the neglect of other things in your life.[17]

Does that sound like your smartphone? No? Where is it? If it's not very close nearby, perhaps it's broken, or you may be one of the lucky few who aren't consumed by their phones. This is the reality of having technology designed for addiction at our fingertips.

An Illuminating Assignment

One of Alter's suggestions is to keep track of your technology use. It's something one of this book's authors has assigned in the form of a media log for more than 15 years. To be honest, students tend to give a collective groan as the media log details are explained. That's because they can't imagine keeping track of all the media content they create and consume over the course of *five* days.

One of the assignment requirements is that students give up some form of media technology for the final two days. And, more often than not, it's either social media or streaming video—things that are more about entertainment than productivity.

A common theme among student reflections after five days of log-keeping is that they know they frequently "disappear" into their phones, but are not motivated to break free from their grip. Here are some representative comments from students, acknowledging the hold that smartphones and apps have on them:

> "I think in YouTube it's very easy to 'go down a rabbit hole' and what I mean by that is you could be watching a video about how to make an instapot recipe, and two hours later you could be watching giraffes fighting each other."

> "I can just pick up my phone and be entertained for hours. It is so much easier than spending time working or learning a skill. The worst part is that I know it's bad for me, but I don't want to give it up."

> "I would like to spend less time on Instagram going forward, but I am not sure that I have the strength to stay away. It is just too easy to open, and way too hard to close."

> "I was without it, or my brain knew I was without it, I was not in a good mood. I was irritable for most of the two days and whenever I looked at my phone. I always stayed normal in front of others but I was struggling inside. I was anxious pretty much the whole time. I was anxious something bad had happened that I needed to see, someone posted a bad photo of me, someone needed to contact me, or all three."[18]

While behavioral addictions don't have to be debilitating, they can still have an impact on our lives and personal relationships. Do you sleep with your smartphone at your bedside? Of course you do—it's probably your alarm clock. But do you also respond to messages or check social media in the middle of the night?

It's hard to remember that social media, smartphones, and apps are a pretty recent development, in the grand scheme of things. And it's increasingly evident that there are some negative consequences to all that *convenience*: body dysmorphia from image filters and social isolation, to name but two.

In completing their media log reflections, students have increasingly noted that they feel like their endless attention to their phones has created a generation of introverts. A couple of responses capture that feeling:

17   Alter (2017), p. 22.
18   Student media log reflections, Wiesinger.

"Technology fulfills a basic need to be social without needing to address people in person."

"I have never not been used to seeing people with a screen in front of their face. I feel like I grew up *with* the media."

"My generation was basically like the guinea pigs for social media. That is why part of me hates social media. We were really the last generation to experience childhood without it. It's like eating junk food, you know it can't be good for you, but it tastes good going down."[19]

If that's the case, we shouldn't be surprised. Researchers have long raised concerns about the technology's ability to separate us, even as it gives us unprecedented access to one another. A 1998 study termed this the **internet paradox**, or the idea that increasing digital engagement is associated with declining communication with family members, a shrinking social circle, and increases in depression and loneliness.[20] The research team from Carnegie Mellon University concluded its analysis with what, in hindsight, is a prescient warning:

> The negative effects of Internet use that we have documented here are not inevitable. Technologies are not immutable, especially not computing ones. Their effects will be shaped by how they are constructed by engineers, how they are deployed by service providers, and how they are used by consumers.[21]

In terms of evolving digital construction, many of the apps we use every day have built in **gamification**—gamified communication—to keep us coming back. Gamification means that the addictive qualities of game play, like log-in gifts and engagement streaks, give us added incentive to keep opening that app.

Duolingo, a very popular language-learning app, is a great example of gamification, with daily, weekly, and monthly user competitions focused on accumulation of points, as well as a limited number of "lives" if you make too many mistakes in a lesson. While Duolingo is actually encouraging you to spend your time learning something, it's also a colossal time suck if you're competitive enough to want to stay in a particular league.

As with many things about technology design and use, gamification isn't inherently bad. As Alter noted, gamification "infuses mundane or unpleasant experiences with a measure of joy,"[22] even as the developer provides incentives designed to be irresistible to the user.

Alter has a few suggestions for breaking your smartphone's hold on your life:

—   Cordon off the technology with self-imposed time limits, like not going on social media after 11 p.m. or before 9 a.m.

—   Go outside and absorb reality, sans device.

—   Have a face-to-face conversation.

—   Stow your device when you're with others, particularly during meals.

—   Create a space with *no* devices.[23]

---

19   Ibid.
20   Kraut, Patterson, Lundmark, Keisler, Mukopadhyay and Scherlis (1998).
21   Ibid., p. 1030.
22   Alter (2017), p. 316.
23   Dreifus (2017), op. cit.

Upon taking a self-imposed tech break after three days of media log data collection, one student pointed to the benefits of a brief tech cleanse:

> I found myself having so much more time in the day to do the little things like dust the windowsills and clean the doorknobs. I was able to sleep better and, when I woke up, I did not immediately reach for my phone as I usually would.[24]

Life in the Age of the Extended Mind

Beyond the addictive qualities of the phones and apps we use every day, studies are starting to unpack the effects they have on our brains.

Some argue that our ubiquitous digital engagement is destroying our memories. For example, studies have shown that frequent use of GPS may contribute to shrinking of the hippocampus—the part of your brain that helps you navigate the physical world.[25]

Others counter that not memorizing things like phone numbers or having to read maps opens up pathways for creativity and innovation.

At the core of both arguments is the concept of **memory**, which the book *Memory and Technology*[26] defines as "information transmitted across time"—knowledge and experience we can draw from, consider, and apply to in the present.

The authors pointed to the difference between **internal memory**, info you store in your brain, and **external memory**, info that gets stored elsewhere.

They identified two basic forms of external memory: social, info stored in other people's memories; and technological, info stored low-tech repositories like books (e.g., encyclopedias), and high-tech resources like the World Wide Web.

In short, we can effectively offload particular forms of internal memory—semantic and prospective memory—by partnering with digital technology. **Semantic memory** helps us remember things like phone numbers, song lyrics, passwords, and trivia, while **prospective memory** reminds us of things happening in the future.[27]

Today, your smartphone provides handy external tools to do both kinds of cognitive work. If you put someone's phone number into your phone, that's semantic memory. If you're setting timers, alarms, and calendar reminders, that's prospective memory.

Once either form of information has been converted to and/or is accessible as digital information, we may never commit it to internal memory. You may be old enough to remember memorizing your home landline phone number, which you still may remember decades later. Now, think of that phone in your pocket: How many phone numbers do you have memorized?

In 1998, researchers Andy Clark and David Chalmers[28] theorized that our technology use creates a coupled system between our brains and the machine. This works particularly well when the external memory device becomes portable, as with smartphones.

Critics point out that things that can be coupled, can also be decoupled—that overreliance on technology inhibits our ability to complete tasks when we don't have access to that external memory.

---

24   Student media log reflections, Wiesinger.
25   O'Connor (2019).
26   Finley, Naaz, and Goh (2019), p. 6
27   "Is technology destroying our memory?" by S.C. Stuart, *PC Magazine*, which features an interview with *Memory and Technology* co-author Jason Finley.
28   Clark & Chalmers, pp. 7–19.

But Clark and Chalmers disagree:

> These systems cannot be impugned simply on the basis of the danger of discrete damage, loss, or malfunction, or because of any occasional decoupling: the biological brain is in similar danger, and occasionally loses capacities temporarily in episodes of sleep, intoxication, and emotion.[29]

In arguing that our engagement with technology is a two-way interaction, Clark and Chalmers noted that the external memory source can actively influence our beliefs:

> There is nothing sacred about skull and skin. What makes some information count as a belief is the role it plays, and there is no reason why the relevant role can be played only from inside the body.[30]

This is an interesting point to consider, given that computer algorithms are part of external memory. We provide data, the algorithm finds similar, supportive data, which has the ability to confirm, alter, and deepen beliefs.

Some 20 years beyond Clark and Chalmers, researchers accept and actively apply their "extended mind thesis" work to today's technology, which is inextricably woven into our daily lives. Benjamin Curtis focused on the role that Google has played, noting that "our minds literally lie partly on Google's servers."[31]

His argument is clear, but goes well beyond Google. Our reliance on phone tools, apps, digital assistants, and more is influenced heavily by the Frightful Five—Alphabet, Amazon, Apple, Meta, and Microsoft—virtually all of which use artificial intelligence to intuit and respond to our needs. Curtis concluded that to break this integration, we need to reclaim our minds from the machine.

While it's highly unlikely that we're all going to ditch the digital resources we rely upon, there's another tool: Digital literacy. This book, and others like it, are building your awareness that transformation is happening, and may even allow you to use some of that unused brain space to spark creative solutions for counteracting the more negative effects of digital engagement.

Summary

Digital living is a hallmark of the post-information age. To our understanding, it's all about *me;* I am in control and I determine what information pathways I follow and when I follow them. Digital living is composed of partnership, participation, and personalization. The partnership is between us, as users, and Platform Service Providers, which provide us space to participate in exchange for demographics and digital privacy. Personalization comes in the individualization of our pages and feeds, our choice of whose content we want to see and who can see ours, and by receiving hyper-personalized ads. But digital living can have the unintended side effects of making purely digital interactions feel "real," making them feel better—less messy—than real life, and making us feel like we are actually accomplishing something by tallying digital friends and followers. But when our days are clogged with text messages, social media feeds, and dozens (if not hundreds) of emails, what are we actually accomplishing? Are we building and maintaining real relationships or digital replicas? And are we even aware of the algorithms and network structures that are playing such a significant role in how these relationships—both real and replicas—come to make up our social lives?

---

29 Ibid., p. 11.
30 Ibid., p. 14.
31 "Google isn't just a search engine—it's a literal extension of our mind" by Benjamin Curtis, Nieman Lab.

Key concepts

Algorithms
Artificial Crocodile Effect
Asynchronous communication
Behavioral addiction
Daily Me
Digital living
Digital
Disneyland Effect
Electronic age
Enculturation
External memory
Gamification
Information age
Internal memory
Internet paradox
Leap second
Memory
Narrowcasting
Participation
Partnership
Pavlov's dog
Personalization
Popularity Effect
Post-information age
Prospective memory
Psychic secretion
Semantic memory
Synchronous communication
Semantic Web
Time shifting

References

Alter, A. (2017). *Irresistible: The rise of addictive technology and the business of keeping us hooked*. New York, NY: Penguin Press.

Clark, A. & Chalmers, D. (1998, January). The extended mind. *The Analysis Committee, Oxford University Press, 58*(1), 7–19.

Curtis, B. (2018, September 4). Google isn't just a search engine—it's a literal extension of our mind. *Nieman Lab*. Retrieved from: https://www.niemanlab.org/2018/09/google-isnt-just-a-search-engine-its-a-literal-extension-of-our-mind/

Dreifus, C. (2017, March 6). Why we can't look away from our screens. *The New York Times*. Retrieved from: https://www.nytimes.com/2017/03/06/science/technology-addiction-irresistible-by-adam-alter.html

Finley, J. R., Naaz, F., & Goh, F. W. (2019). *Memory and technology: How we use information in the brain and the world*. Switzerland: Springer Nature.

Huszár, F., Ktena, S. I., O'Brien, C. & Hardt, M. (2021) Algorithmic amplification of politics on Twitter. *Proceedings of the National Academy of Sciences, 119*(1).

Kraut, R., Patterson, M., Lundmark, V., Keisler, S., Mukopadhyay, T., & Scherlis, W. (1998). Internet paradox: A social technology that reduces social involvement and psychological well-being? *American Psychologist, 53*(9), 1017–1031.

Mandell, A. (2014, October 1). *Movie stars grapple with family life in the digital age*. USA Weekend.

Negroponte, N. (1995). *Being digital*. New York, NY: Vintage Books.

Obleukhov, O. & Byagowi, A. (2022, July 25). It's time to leave the leap second in the past. *Meta Production Engineering*. Retrieved from https://engineering.fb.com/2022/07/25/production-engineering/its-time-to-leave-the-leap-second-in-the-past/

O'Connor, M. R. (2019). *Wayfinding: The science and mystery of how humans navigate the world.* New York, NY: St. Martin's Press.

Riley v. California, 13–132, U.S. (2014). *Supreme Court of the United States.* Retrieved from https://casetext.com/case/riley-v-cal-united-states-1

Stuart, S. C. (2019, June 5). Is technology destroying our memory? *PC Magazine.* Retrieved from: https://www.pcmag.com/news/is-technology-destroying-our-memory

Turkle, S. (1995). *Life on the screen: Identity in the age of the Internet.* New York, NY: Simon & Schuster.

Vincent, J. (2022, July 26). Leap seconds cause chaos for computers—Meta wants to get rid of them. *The Verge.* Retrieved from https://www.theverge.com/2022/7/26/23278718/leap-second-computer-chaos-meta-backs-campaign-to-end-it

CHAPTER 6

Digital Identity: Options, Opportunities, Oppressions, Impressions

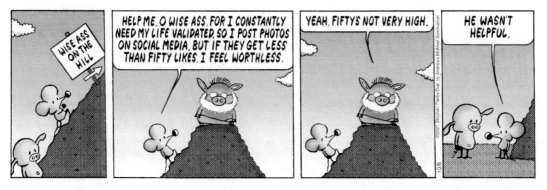

Figure 6.1. The quest for validation online almost inherently breeds challenges to self-worth. Pearls Before Swine © 2021 Stephan Pastis. Reprinted by permission of Andrews McMeel Syndication. All rights reserved.

Takeaway: There is a relationship between who we are online and who we are in the real world, but that relationship is more complex than we might at first think. And it changes.

What's your social media user name?
What's your profile picture?
How about your cover photo?
Which one?

If the latter is your first thought upon reading the preceding three questions, you're tapping into the issue of multiple and overlapping identities—a hallmark of digital living.

Perhaps a better question is, "How do my social media postings reflect who I am right now?"

Consider for a moment the way you—and people around you—participate in social media: pictures of pets, pictures of food you are about to eat, pictures of yourself, pictures of the weather, a funny story, a complaint about a store, an opinion about a movie or a TV show or a band.

All of these little fragments represent moments that are unique, but in an odd way they seem to be the same moments that everyone else is having.

The reproduced moments have meaning that is directly related to who *we* are, but sometimes the consistency with how others represent themselves online is notable.

Look at the pictures of yourself you have put together over time on social media—your selfies. What you see is an image of yourself added to a whole timeline of other images of yourself. The person, pose,[1] and surroundings might be the same, but the mood and corresponding identity have shifted with each image.

As you took and uploaded each image, they were shared with the people you've selected for your social network.

Our social media posts are evidence to others that we *are*.

We are part of a collectively constructed reality.

We are worthy of the attention of others.

And, yes, we are incredibly conscious of how we are perceived online. We make choices about when and how to negotiate our identities in interactions both online and face-to-face.

It would be one thing if the two were separate, if there were no people seeing the relationship between the two. But we have various levels of interaction with people online, some who maintain physical contact with our present time and space, some who are never in close contact, others who *were* in contact but now are more distant. And there are still others who we have never once been in contact with.

It's interesting to have acquaintances and friends who deliberately alter who they are online, when we are close enough to see the difference. Perhaps their online identities are parceled out, divided, and restricted from their physical-world identities.

Which "You" Am I Looking At Here?

Sometimes seemingly odd online deviations from what we think we know about a person are not intentional.

That's because the moment-to-moment "realities" of social media postings collect into a timeline that can—and can't—be reflective of a person's physical life.

---

1   Do you have a preferred selfie pose? Take a look at your feeds and examine the pictures of friends. Bet you'll see that most have a go-to posture and facial expression. Once you see it, you can't unsee it.

While the images themselves are from a fixed point in time, their place on the social media time-line is remarkably fluid. Posts from the past—some only a few minutes, other stretching back years—are given currency by their place in another user's feed.

So instead of a fixed plotline that reflects the chronological order of a person's online life, we get variations of identity and reality from the past that can change who we understand someone else to be *now*.

Maybe an example would help here.

Someone you knew in elementary school tracks you down and wants to be friends on Facebook. You accept. Part of the ritual at this point is **feed fishing,** or getting to know another person by looking at his or her social media feed—particularly photos. In the case of someone you knew in the physical world in the past, feed fishing helps you align the person you see online today with the person you remember.

Say you run across a photo from five years ago that shows a tattoo in progress on the person's back. You "like" it and comment, "Great tattoo!" That effort pulls the tattoo picture back into the other person's current feed—out of time and context. Others note the "new" tattoo and comment on it.

Rather than holding its place in the individual's physical timeline—marking an event that took place five years ago—the tattoo picture represents current reality for those who see it (again) pop up in their feeds.

And current comments on the five-year-old post, like "How cool! When is it going to be finished?" and "I didn't know you had that," may take on a surreal quality, even for the person who originally posted the photo. The tattoo we're commenting on as in progress *now* was completed *five years ago*.

The Cult of Personality and Anonymity

Issues of identity online are further complicated by the fact that we may divide ourselves into different "users," depending on who we think is looking.

Sometimes people use their own names, but sometimes writers adopt a literary **doppelgänger**—an alter ego—to separate their various views from each other and from other parts of their lives.

For example, Alexander Hardy is an insightful writer who posts his work at different locations across the web and via social media. One nom de plume—pen name—is "the colored boy."[2] This is where Hardy puts some of his most provocative work on race and inequality. In this instance, Hardy is hiding in plain sight, as he identifies himself as the blog's author in its "about me" section.

The nature of the web and social media is that we *can* maintain a high degree of digital anonymity, which allows us to promote and defend our views and values online, without having to answer for—or even acknowledge—strident posts and personal attacks in our physical lives.

There are complications and benefits to anonymity. It allows you to interact with someone online without allowing them to know your true identity. But anonymity also has a dark side: if we *know* we can't be identified, if we *know* our words can't come back to us in the "real" world, we may be less inclined to engage in civil discourse.

For example, a Twitter user created an account called Devin Nunes' cow (@DevinCow) specifically to troll—in this case heckle, mock, and make fun of—then-California Congressman Devin Nunes. @DevinCow went to great lengths to hide their identity, and Nunes went to great lengths to discover it.

Nunes sued both @DevinCow and Twitter, seeking $250 million in damages for defamation of character. When the suit was filed, @DevinCow had 1,204 followers. Days later, it had more than half a million.

2   Note the lowercase title—this isn't an error. Our online identities often are managed right down to how we use capitalization and punctuation.

The court determined that Twitter was protected as a responsible host and was removed from the lawsuit. That left the anonymous account, which Nunes relentlessly pursued. Although Nunes left congress, @DevinCow remained active on Twitter, peaking at about 771,000 followers.[3]

We also can build and maintain distinct "brands" that encompass our professional aspirations, family, friends, acquaintances, and hobbies. And not only do we have the ability to simultaneously maintain several digital identities, we can even have them talk to each other if we so choose.

When two (or more) digital identities created by the same person communicate online, they're known as **sock puppets.**

While that tactic might sound a little strange, sock puppets frequently are used to shore up ideologies and arguments in commenting, as well as adjust the direction of posts on review sites (e.g., Yelp and Tripadvisor). That makes the reasoning understandable, but it also makes the use of sock puppets somewhat unethical.

Case in point: U.S. Senator Mitt Romney was unmasked in 2019 as the author of an anonymous Twitter account that defended him and criticized then-president Donald Trump—notably in response to Trump's direct attacks upon him, including the hashtag #IMPEACHMITTROMNEY.

The senator admitted to being Twitter user Pierre Delecto (@qaws9876) after a journalist did some simple sleuthing and called him on it. The account launched in 2011, when he initially started his run for president, and his first followers were his eldest son and trusted advisers.[4]

Pierre Delecto defending Romney's honor is definitely sockpuppetry.

Online Personalities & Personas

Keep in mind that the communities and individuals we interact with online are a complex combination of what we experience in our face-to-face world and the effects of the scaled-up network of social media connections.

Among the people we interact with online are **influencers**, people who use their online personas to guide the actions of the people who frequent their feeds. Sometimes this brings products to the awareness of an audience, as with the case of 2022's viral Pink Sauce. Influencers will receive products in the mail for them to use and, hopefully, promote through their highly influential channels.

Sometimes the influencer-brand partnership is obvious and ad-like, and sometimes it's a more of a consistent brand presence, as with the sponsorship of Yoga With Adriene by Adidas. Other times, influencer sponsorships exist as **sponsored content**, where brands pay for prominent use and display of their products. Whichever way the sponsorship evolves, the influencer is compensated in one way or another.

For example, travel influencer Christine Amorose Merrill, who goes by the handle C'est Christine (@cestchristine) across all channels, frequently promotes brands in her travel posts by showing them. She encourages brand contact and acknowledges brand sponsorship with this statement:

> Full disclosure: I do accept sponsored content (because if I didn't, I wouldn't be able to travel as much and do nearly as many fun things to write about), but all opinions are my own. I'm passionate about only promoting goods and services that I truly believe in and use myself.[5]

Influencers operate at different levels, depending on the size of their following—spanning from those with a thousand or two viewers that reach a desirable niche market, to celebrity influencers who can attract audiences in the millions.

---

3   "Devin Nunes can't sue Twitter over cow and mom parodies, judge says" by Bryan Pietsch, *The New York Times*.
4   "This sure looks like Mitt Romney's secret Twitter account" by Ashley Feinberg, *Slate*.
5   https://www.cestchristine.com/contact/

Getting Verified and Authenticated

Several social media sites have developed means of influencer **verification**, which signals that the account has been authenticated by the site. It might be a checkmark, or a crown, or some other indication that posts are being done by the person who is certified to be who they claim to be. This process helps followers to know that the person they *think* is influencing them is actually that person—not tribute accounts, fan accounts, or fake accounts.

Verification confirms the *source* of the postings. It's not necessarily an indicator of their status, qualification, or accuracy.

What Do Influencers, Well, Influence?

Influencers have developed a reputation that viewers value for their expertise and experience on a specific topic. They post on a regular basis and interact with topics in a way that followers are attracted to.

They essentially become a branded demonstration for a specific social or cultural pursuit. They might be a person who reviews books, and leads others to buy and read what they recommend. Or they might be a kid who has followers because of the way they unbox new toys.

The influence comes in whether or not the follower chooses to adopt a product or activity based on influencer recommendation.

Some of the most popular influencers—and some of the biggest earners—are celebrities who work with marketing agencies to link products with their "brand." They come from the worlds of sports, or music, or television, or purely through their celebrity presence. Some celebrity endorsers can earn well over a million dollars for a single sponsored social media post.

Corporate sponsors can pay influencers for talking up their products, as well as for wearing them, using them, or consuming them. The success of this is directly tied to the size and enthusiasm of the audience of followers.

Working with influencer marketing agencies, a brand will pay an influencer a fee for a spot or if it is a campaign perhaps for series of spots. Some influencers may even get a percentage of sales for brands with whom they have a relationship.

Other influencers have opportunities to get contributions from their followers, using things like virtual tip jars. This has been a feature of Twitch for some time, where players with followers set up donation links through third-party tools—allowing people to reward them during their play. Other services like Twitter and Soundcloud have developed similar ways for listeners and viewers to tip the people they follow.

Memberships

While some influencers thrive on YouTube, Instagram, or TikTok channels, others can be found in the podcasting world, where again their expertise and persona generates listeners. Those listeners can then pay for a membership to someone they follow, allowing the people followed to get compensated for their work.

Patreon is one such service, working something like a subscription service linking creators with fan communities. Often these memberships allow followers access to content that is exclusively available for members of the fan community. This is known as **crowdfunding**, which generates income for creators from the crowds that follow them.

The success of crowdfunding, or generating subscriptions, or getting tips, or getting sponsorships is tied to the personality of the influencer. When we think about the authenticity of people we follow, we tend to be conscious of how we are attracted to their "realness." In reality, that authenticity may well be an identity that the creator or influencer has spent time and effort crafting and refining.

Is this who or what they really are? Or is it a constructed identity? In most cases, as followers, we may never know. And for some creators—as you might expect—their success influences who and what they really are. As Kurt Vonnegut wrote, "we are who we pretend to be, so we must be careful what we pretend to be."[6]

Negotiating Identity

This notion of having multiple and conflicting identities isn't really new; it's just different because of our present ability to sometimes see two different people—the physical and digital selves—at the same time. Maybe people shape who they are because they know that parents are watching or because they are trying to maintain a boundary between themselves and their presentation for future job prospects or current coworkers.

It would be useful to think for a moment about how this works, starting with drama. We know we like a little of it; we know the cost of too much of it. But when we think of ourselves in the juxtaposition of real and digital worlds, a drama metaphor has been used—and we should find it quite useful.

What are the elements of drama? We have performers, we have audiences, and we have a play.

Online you are an actor performing a version of yourself when you are adding elements—stories, pictures, videos, and so on—to the character that is you. And when you are watching the characters performed by others, you become an audience, watching the unfolding drama of the lives of others. Even when we think we are being ourselves, what we are doing can be thought of as a performance.

You wouldn't be the first to think through the metaphor of drama for understanding our identities. Sociologist Erving Goffman developed the theory of **dramaturgy** more than half a century ago to describe how our lives function as performances.

Pretty much right from the start, Goffman was interested in understanding how people control the presentation of themselves. He was particularly interested in the framework within which an individual works to maintain multiple and conflicting identities:

> He may wish them to think highly of him, or to think that he thinks highly of them, or to perceive how in fact he feels toward them, or to obtain no clear-cut impression; he may wish to ensure sufficient harmony so that the interaction can be sustained, or to defraud, get rid of, confuse, mislead, antagonize, or insult them. Regardless of the particular objective which the individual has in mind and of his motives for having this objective, it will be in his interests to control the conduct of the others, especially their responsive treatment of him.[7]

Does this seem familiar? Are these the kinds of thoughts that cross your mind as you interact with people, especially online?

There is much to consider in Goffman's work, but we want to pay particular attention to how he sees us as actors on the stage of our lives. He sets out the terms **front stage** and **back stage** as ways of looking at the way we interact.

The front stage is a place where we essentially try to behave in a way that observes politeness and decorum. We try to fit in, not offend others, and generally get along.

Our understanding of the rules of the front stage area is quite sophisticated and anchored in the culture in which we live. Sometimes we play with the rules of this front stage area, and for some people we tolerate violations of the rules.

Perhaps this could be best embodied in considering what is OK and not OK to talk about to other people in public. Many people shy away from complicated topics, like, say, politics and religion. But whether a particular topic area is appropriate really depends on how we understand the nature of front stage.

The back stage area comprises those things that we don't think are part of our public presentation. Maybe we think these are not anyone else's business. Maybe we think of it as a kind of privileged

---

6   In *Mother Night* (1961).
7   Goffman (1959), p. 3.

communication that deserves a greater measure of privacy. Maybe it's a place where we just cut loose, let our hair down, and suspend the restrictions of politeness and decorum.

But it's also the place where we can talk about some of the more restricted topics—what we believe in spiritually and politically, how much money we make, what kind of intimacy we desire and enjoy. This is probably the place where guilty pleasures are enjoyed or perhaps where our more base prejudices are expressed.

All this stuff is behind the curtain, because we don't necessarily want to be public about the way we look without our masks on or what we say when we are not inhibited by being in performance mode.

Now think of the way that people draw this line when they make decisions about what to post online. Some people are clearly very guarded, careful about what they put online. Needless to say, we need to extend this way of thinking to both words and images.

And keep in mind how frequently we don't even think very carefully about this difference. Perhaps you have posted something in haste to only later find yourself regretting that you posted it. Sometimes you can edit or delete it, but we need to remember it has already gone out into a space that does not like to be corrected. And we have to keep in mind that many of these social platforms have as a feature the ability to re-post things with the greatest of ease.

Sometimes the machinery of online and social media has some of these protections built into their technology. You can make choices about who is in your circle and how much of your life they are privileged to see.

But that curtain between front stage and back stage is something that has gone through—and continues to go through—transformation related directly to media technologies.

Identity Boundaries Are More Complex Online

Goffman's idea was that we all have parts of ourselves we put on display and parts we keep away from the audience. Online complicates the notion of front stage and back stage behaviors, however.

Think for a moment about how something as simple as your birthday plays out online. Some people may choose to keep birthdays relatively private to avoid the attention. Others may celebrate them in public. Each of these options reflects a degree of individual choice as to whether a birthday is a back stage or front stage performance.

But there's a third option that arises because we give social media platforms permission to post for us. It plays out via the social media notifications that alert friends and followers that it's your birthday.

Sometimes this comes across as a mini surprise party, as birthday bait sent to social media "friends" and email connections often leads to a lengthy chorus of "Happy Birthday!" in social media feeds. But for others, that forced interaction feels like the birthday equivalent of writing "Have a good summer!" on a yearbook page—you sometimes do it because you're obligated and don't have anything else to say.

Author Joanne McNeil characterized birthday "outing" by social media—taking something she preferred to be back stage and yanking it to the front stage—as **birthday harassment:**

> It was my birthday recently. Perhaps you heard? Sorry about that! Google Plus, the zombie social network I have barely used since its launch in 2011, alerted my contacts that have Android phones. And anyone with iCal synced to Google Calendar had it marked in their iPhones. "The internet is trying to tell me that it's your birthday," someone wrote in an email to me. Throughout the day several others would send similarly kind but bewildered messages. [8]

McNeil was pointing to the interconnectedness of our electronic worlds. And while she called the experience "embarrassing," her greater concern was the underlying reason that Google had co-opted her identity.

---

8   "The internet of things will ruin birthdays" by Joanne McNeil, Medium.

Google's strategy was to send the mass alert to her contacts, then scour her Gmail account for messages from those contacts that contained the word "birthday." Google then could give those contacts a higher **affinity ranking,** which means that they likely are people who influence McNeil or are influenced by her. That ranked network then could be marketed to advertisers or used in yet-to-be disclosed research projects initiated by Platform Service Providers.

McNeil acknowledged that her own usage patterns might confound the machines and confuse her friends: She sometimes picks January 1 as her birthday when it comes up as the first option, because it's easier than clicking on a drop-down menu and scrolling to her real birth date.

What is the upshot? The construction of our online identity isn't always a conscious choice, with parts of that identity co-opted and controlled by how others respond to auto postings from computer algorithms.

Constructions of identity also carry with them some of the frictions and conflicts that exist in the culture that produces them—including reproducing things like racism and sexism online.

Take, for example, the case where an international beauty contest was judged by artificial intelligence—by robots. The creators of the contest, called Beauty AI, expressed dismay that the robots did not like people with dark skin. Sociologist Ruha Benjamin pointed to this preference as a reminder that it was the humans behind the contest who did the original thinking and programming for the robots.[9]

Reproductions of cultural attitudes into the digital sphere spans to companies that run search engines, which pay attention to how their algorithms communicate ideas about identity.

Internet scholar Safiya Noble's work focuses on how search results can reflect problems with race and gender in society.[10] She extends this criticism to look at patterns in online paid advertising, and discusses how private interests and attitudes within the digital industries reproduce patterns of racism and sexism.

Noble pointed to a 2017 internal manifesto from a Google engineer that argued that women are more interested in people than things, which makes them prone to being neurotic and makes them psychologically inferior. That person was fired for violating the Google code of conduct, but Noble argued that this is not unique in the online world.

What we gain from these arguments is a good note of skepticism. Connections between ideas in the online world may be run by and through algorithms and AI bots, but they will still reflect the cultural attitudes of the kinds of people that create and maintain the programming.

This is a good reminder that what we consume online (re)shapes our identities.

Making a Spectacle of Yourself

When we make information part of the digital world, we only have a certain ability to control what meaning is made of that information. Some inside jokes make sense to just a few people, some to a larger group, some to a whole culture. But you have to be aware of the limits.

There are now many true accounts of people on Twitter who thought they were being funny, but they forgot how their tweets might look, and they also forgot how they might get distributed. And sometimes, what was intended to be a back stage behavior, takes a front stage twist that connects digital and physical realities in sometimes self-destructive ways.

Social media shaming is the idea of calling out social media users—often strangers—for insensitive or offensive posts. Sometimes a post intended only for friends and followers goes viral—is shared well beyond user-selected limits—and can have real-world consequences.

9   Benjamin (2019), pp. 49–50.
10 Noble (2018).

The destruction of Justine Sacco's life is a prime example of what "one stupid tweet"[11] can do. In December of 2013, while Sacco was waiting for a flight to South Africa, she made what she thought was a funny post to her 170 Twitter followers before she settled in for an 11-hour plane trip: "Going to Africa. Hope I don't get AIDS. Just kidding. I'm white!"

The post, which later became known as the "tweet heard 'round the world," went viral, spurring tens of thousands of angry responses from Twitter users. Sacco, the then-30-year-old director of corporate communications for the global Internet company IAC, landed in South Africa to discover that she (a) was one of the most-hated social-media users on the planet, and (b) had been fired for what was widely interpreted as a racist, rather than snarky, tweet.

What started as something to fill time in the moment had become Sacco's identity to much of the rest of the world, thanks to hundreds of thousands of shares and widespread media coverage.

"All of a sudden you don't know what you're supposed to do," Sacco told *New York Times* reporter Jon Ronson. "If I don't start making steps to reclaim my identity and remind myself of who I am on a daily basis, then I might lose myself."

Ronson characterized such "shame campaigns" as being a form of "collective fury" that allowed other users to launch en masse attacks on individuals. He pointed to the practice as spanning the distance between physical and online identities, effectively destroying both.

The critical connection here is the idea that front stage and back stage behaviors are blurred online. In fact, the only way a person's actions can truly be considered back stage is to be sure they don't ever end up online—which, in a world filled with smartphones, is increasingly difficult.

Months after Sacco's fateful tweet, her identity and life still were in flux. She noted that the public humiliation had affected every aspect of her life, from family relations to dating. Ronson concluded:

> Social media is so perfectly designed to manipulate our desire for approval, and that is what led to her undoing. Her tormentors were instantly congratulated as they took Sacco down, bit by bit, and so they continued to do so. Their motivation was much the same as Sacco's own—a bid for the attention of strangers—as she milled about Heathrow, hoping to amuse people she couldn't see.

Phenomena like social media shaming become important lessons for how we need to better understand how we are related to what we do online.

Keep in mind that this is not just a question of being "politically correct." It's a question of coming to a fuller understanding of how our external communication reflects on who we are—and who we think we are.

Some people think what they are writing and posting online will be clearly understood as intended—that sarcasm and jokes will be obvious to readers. But that misses a particular aspect of the medium, that viewers see the words and only know the intentions that seem to be apparently there. The more complicated, more nuanced way we use words doesn't necessarily read the same way it sounds.

Some people's online lives seem to be a collection of bragging postures. Other people seem to fill their online lives with tragedies. Consider, for just the present moment, how you define appropriate and inappropriate things to do online. You have to consider this for the present moment because it's constantly changing.

Some people think it's OK to break up with somebody via text. Some people think the announcement of the death of a pet or even of a relative is an inappropriate thing to do on Facebook. But these are changing, and in your social circle they may already be considered completely acceptable behaviors. Just be careful how much you assume everyone shares the appropriateness of certain behaviors.

---

11 "How one stupid tweet blew up Justine Sacco's life," by Jon Ronson, *The New York Times Magazine*.

Your communication is inescapably a product of the history against which your life happens. Becoming more information literate means becoming more sensitive to the historical context of what you're doing both online and in your face-to-face life.

Ultimately, you will choose what you think is appropriate based on interaction between your own values and the values of the groups to which you belong. But that doesn't mean that the only people who will see these things about you will be people who understand.

The Power of Ideology

Ideologies are a web of influence, ideas, and beliefs, that tell us how we should live. In day-to-day life the word ideology, like other complex terms, is used in different ways. We need to be sensitive to how the term is used, and rather than worry about whether the use is correct; to look for the reason that people would want to use the term in a particular way.

Because of the sizable power of ideology, the term is commonly used in a negative way. In this use, people act *ideologically* when they have been "brainwashed" by a set of beliefs, and are no longer acting in a reasonable manner. It becomes an us-versus-them frame: They are wrong because they are acting out of their ideology; we are right because we are acting out of reason, or moral judgment. While this is a very common use, it misses the point that we all operate out of belief systems.

Even if you want to insist on the superiority of your (or, more accurately, your group's) belief system, your ability to communicate to those outside the group is limited. Others may not appreciate the way you are looking down on their belief system, or may be looking down on your group's belief system.

The result is a stalemate, or at least a lack of communication. And your ability to understand across lines of difference is greatly reduced.

What we might want to hang onto from that more negative way of viewing (i.e., it brainwashes you), however, is the idea that *ideology is very powerful*.

We might want to be skeptical and critical about how our ideology—our belief system—can be and is used against us. This is especially true when it gets us to do things and believe things that we don't think or do because they are in our best interests, but just because we think they have always been done that way.

That is the great power of ideology: Our sense is that it is eternal, has always been that way, and is the same for every group and every person. Ideology goes through a process of *naturalization*, which suggests it has power through its stability—it's the way it has to be by definition.[12]

That naturalization starts when people are young and are just beginning to acquire fixed ideas, attitudes, values, and perceptions. So early on we are formed in a specific ideological way, heavily influenced by family and surroundings. And once formed, it is seen as *natural*.

You can hear echoes of this when people say something is "just in a person's nature." It suggests that the belief is inflexible and automatic—a fact rather than a choice. So we forget where ideas came from, and what context gave them their ideological power in history.

Despite the experience of our ideologies as if they were eternal, universal, and unquestioningly correct, we gain from being skeptical, from reflecting on and looking at where our ideas and our beliefs came from and from whom. That reflection and questioning is the road to digital, media, and information literacy.

Again, that does not mean you have to change anything—and you may be part of a group that would find such an investigation threatening. But note what that situation might suggest: An idea or a belief of some kind is sustainable only if you don't look very closely at it, see where it came from, and understand its history. A perspective seeking a deeper literacy ought to become very suspicious under such circumstances.

12   Yewah (1993).

Keep in mind that we think of our cooperation as the product of our own will. We want to fit in, we want to belong, we want to do what is good for the groups we are part of, so going along is seen as the best way of getting along.

The Intersection of Ideology and Digital Identity

Today we have virtual access to an extraordinary range of views about any topic, which *should* broaden our perspectives. Instead, our social media feeds are full of friends and followers who tend to share a similar belief system. Views that challenge our own are easily excluded, ignored, and left unexplored.

This is the classic **echo chamber** effect, which means that we trap ourselves in virtual echo chambers of our own making, shouting out to the world and hearing our own voices and views come right back at us.

Social media are particularly good at creating echo chambers.[13] We choose the online communities we want to be part of, choose our friends, choose our followers, and choose whom to follow. The voices and views that come back at us tend to be very much like our own.

Even if you want to insist on the superiority of your (or, more accurately, your group's) belief system, and even with the global nature of digital media, your ability to communicate to those outside the group is limited. Others may not appreciate the way you are looking down on their belief systems, or they may be looking down on your group's belief system.

The result is a stalemate or at least a lack of communication. And your ability to understand across lines of difference is greatly reduced.

Digital Media as Ouroboros

Social media experiences make many people happy because they feel connected and connected *right now*. Often, they gravitate toward people who share the same values or perhaps even the same opinions. There is a comfort in that as well, since it works against feeling isolated and alone. But the nature of platforms like the web and social media is that we tend to produce content for others who share our views, who in return feed our views back to us, which we comfortably consume.

This cycle calls to mind the image of a snake that is eating its own tail. It's called an **ouroboros**, and it is often depicted like this:

With this image, which has been used since ancient times, we can think about the way things cycle—seasons, parts of our lives, and especially cultures remaking themselves. Remember that ideologies have to be reproduced in each new member of the group.

Figure 6.2. And this is what the ouroboros looks like. Just as we said: A snake eating its tail.

---

13   A 2015 study by Facebook, however, found that the echo chamber effect is not a result of the social media giant's underpinning algorithms, which are designed to optimize content for users. Instead, Facebook researchers found that political biases that crop up in a news feed are the result of who and what the user has decided to like and follow. In short, the classic echo chamber effect (Rundle, 2015)

Where the idea of the echo chamber is almost always negative, since it suggests repetition and volume replacing reasoned consideration, the snake eating its own tail is a more complicated idea. Is it self-destructive, or is it complete?

When you spend time online you might find yourself sinking into repetition and reinforcement—attending only to attitudes, ideas, and beliefs ideas that are like yours, blocking, muting, and/or unfollowing anyone who disagrees or challenges. That's the ouroboros: We're comfortably consuming ourselves under the active influence of the companies that program the platforms we engage with every day.

Misinformation and Disinformation

Social media, in tandem with our ever-present smartphones, acts through a kind of constant hailing—reaching out to you, 24/7, seeking constant attention. It feels good to be connected, to be socially involved, to interact with our social networks.

You have heard this before, but it bears repeating: There are lots of differences between face-to-face interactions and interacting through technology. But a big part of many cultural ideologies is that technology is good. And it is good for you. And it gives you a constant source of pleasure and perhaps even self-esteem.

Think of the idle moments when you turn to a device and check up or check in, or interact, or wander online. We are busy, involved, engaged, and happy to feel alive in the technology.

But continuous exposure, coupled with the pace at which information comes at us, means we have very little time to consider the veracity of what we're consuming. There's also a tendency to question information less if it comes from a trusted source—and trusted sources on social media are people we've chosen (again with the ouroboros!).

There's a thought to consider here, which is a kind of old saying. It goes like this: "Everyone is entitled to their own opinion, but not their own facts." This has been attributed and often misattributed, but the idea has merit. It states the importance of your individual opinion, but warns against basing it on beliefs that are not grounded in shared sets of facts.

Facts, in fact, are a slippery thing. Our understanding of things like science, evidence, and confirmation are increasingly evaluated via the filter bubbles we construct on social media.

Filter bubbles allow us to exclude thoughts that challenge our ideas and beliefs. It's a positive feeling to consume perspectives that align with our own, giving us a sense of community and belonging. But what we don't notice as much are the limitations on our freedom of choice that are part of the effect, made more complicated by the algorithms that help shape our filter bubbles and the way it all operates somewhat invisibly.[14]

And, if we're not carefully considering the information that comes from our selected sources in our selected feeds, we're more vulnerable to things like misinformation and disinformation.

Misinformation is false information that is spread through legacy or digital media that is created and spread regardless of any intention to harm or deceive. **Disinformation**, on the other hand, is intentionally false and designed to deceive, mislead, or confuse.

Among the many challenges of living in a media world that presents a lot of misinformation and disinformation is the problem that false information spreads much faster and much more widely than true information.[15] According to a 2019 study from the Massachusetts Institute of Technology, false information is 70% more likely to be retweeted on Twitter than the truth, and reach the first 1,500 people six times faster.[16]

---

14 "Breaking the filter bubble: Democracy and design" by Engin Bozdag & Jeroen Hoven, *Ethics and Information Technology*.
15 "Rumor's reasons" by Farhad Manjoo, *The New York Times*.
16 "What's the attraction to the spread of disinformation?" by Fathia Eldakhakhny, *Center for Media and Peace Initiatives*.

And once misinformation begins to spread, it's hard to stop.

For example, during the great Texas freeze out of 2021, Reza Aslan, a scholar with nearly 300,000 followers on Twitter, retweeted a tweet from Texas Senator Ted Cruz, saying "I'll believe in climate change when Texas freezes over." As they say, there's a tweet for everything—and this one, dated 2016, did not age well.

Moments later, Aslan realized the tweet was a hoax, deleted it, and corrected his mistake with another tweet, "Oh, I guess it's not."

Aslan's original tweet had 117,000 likes and nearly 25,000 retweets. The correcting tweet received 522 likes and 100 retweets. In short, the lie spread, the truth was ignored.

SIFT and the "Four Moves"

There's a swirl of complications around this issue that affects each of us. In basic terms, the more we invest in our own media and digital literacy, the better we are at detecting misinformation and disinformation.

Sometimes this means double checking something that we read from a source that we don't know very well, and checking to see its reliability. The SIFT method says: **Stop**: **Investigate** the source: **Find** trusted coverage: and **Trace** to the origin.[17]

This can often be done very quickly, in a process that was outlined by education-technology expert Mike Caulfield,[18] who suggests that we can use a few simple digital checks to see if what we are seeing is misinformation or disinformation.

These steps are:

1. Check for previous work
2. Go upstream to the source
3. Read laterally
4. Circle back

We **check for previous work** by looking at fact checking websites, for example, to see if the item in mind has been fact-checked. Example sites include Snopes, PolitiFact, Factcheck.org, NPR Fact Check, and Hoax Slayer. You can find these sites pretty easily, but you still have to be responsible for knowing what that site does and how it works.

Going upstream means seeing where the information came from, going to the original source. This can also require that you understand how news and information works in real time. Many rumors and hoaxes arise out of news coverage of a breaking story that reports a bit of information it later retracts.

Reading laterally means to go out of the site where you found the information and look at how others are covering it. It works in tandem with going upstream, since the idea is that multiple places offer more solid confirmation.

Finally, **circling back** is a good strategy if the checking process gets you lost, or worse, following down a rabbit hole. You can go back to where you started with the new information you gained from the first run at checking, and find other search terms or other places on the web that can confirm or refute the idea.

---

17 Remember librarians? You can see the kind of assistance they offer on a subject like this; see University of Oregon Libraries, https://researchguides.uoregon.edu/fakenews/sift.

18 Agee (2018), Caulfield (2021).

One other part of this is critical: taking a pause. It is often the case that we have strong emotional reactions to a story or a piece of new information, and we are compelled to share or re-tweet it. This approach strongly suggests waiting—taking a step back from your strong emotional reactions.

Much of this works well when you are feeling patient, when your identity is not being challenged, or when you're in a frame of mind that is OK with being wrong. The worst position to be in is when you are so convinced of the truth of something, that hanging onto that belief is more important than all of the evidence that refutes it.

Summary

We live divided lives, fostered by a culture of digital feeds, friends, and followers. This allows us to create multiple and overlapping identities online, which may or may not correspond with the views and values we share in our personal lives. We have the ability to build and maintain distinct "brands" that encompass our professional lives, relationships with family and friends, and various hobbies and personal interests. Sometimes things that we think of as back stage behavior (i.e., acceptable for some audiences but not others), unexpectedly cross into the front stage of our digital and/or physical lives. At the same time, our past and present are muddied by a social media plotline that is not fixed in either time or space. This requires us to constantly adjust and negotiate our online identities, as well as how they reflect on our physical lives and interact with the beliefs and identities of others.

Key concepts

Affinity ranking
Back stage
Birthday harassment
Crowdfunding
Disinformation
Dopplegänger
Dramaturgy
Echo chamber
Feed fishing
Front stage
Influencers
Misinformation
Naturalization
Ouroboros
SIFT
Social media shaming
Sock puppets
Sponsored content
Verification

References

Agee, A. (2018). Fake news and the Caufield Technique. *Presented at the California Academic & Research Libraries 2018 Conference.* Retrieved from: http://conf2018.carl-acrl.org/wp-content/uploads/2018/08/agee_engaging_fakenews.pdf

Benjamin, R. (2019). *Race after technology: The new Jim* code. Medford, MA: Polity Press.

Bozdag, E. & van den Hoven, J. (2015). Breaking the filter bubble: Democracy and design. *Ethics and Information Technology,* *17*(4): 249–265.

Caulfield, M. A. (2021). *Web literacy for student fact checkers.* PressBooks. Retrieved from: https://webliteracy.pressbooks.com/front-matter/web-strategies-for-student-fact-checkers/

Eldakhakhny, F. (2020). What's the attraction to the spread of disinformation? *Center for Media and Peace Initiatives*. Retrieved from: https://www.mycmpi.org/whats-the-attraction-to-the-spread-of-disinformation/

Feinberg, A. (2019, October 20). This sure looks like Mitt Romney's secret Twitter account. *Slate*. Retrieved from https://slate.com/news-and-politics/2019/10/mitt-romney-has-a-secret-twitter-account-and-it-sure-looks-like-its-this-one.html

Goffman, E. (1959). *The presentation of self in everyday life*. New York, NY: Doubleday.

Manjoo, F. (2008, March 16). Rumor's reasons. *The New York Times*. Retrieved from https://www.nytimes.com/2008/03/16/magazine/16wwln-idealab-t.html

McNeil, J. (2014, July 28). The Internet of things will ruin birthdays. *Medium*. Retrieved from https://medium.com/message/the-internet-of-things-will-ruin-birthdays-8a5b781abb6b

Noble, S. U. (2018). *Algorithms of oppression: How search engines reinforce racism*. New York, NY: NYU Press.

Pietsch, B. (2020, June 25). Devin Nunes can't sue Twitter over cow and mom parodies, judge says. *The New York Times*. Retrieved from https://www.nytimes.com/2020/06/25/us/politics/devin-nunes-cow-tweets.html

Ronson, J. (2015, February 12). How one stupid tweet blue up Justine Sacco's life. *The New York Times Magazine*. Retrieved from http://www.nytimes.com/2015/02/15/magazine/how-one-stupid-tweet-ruined-justine-saccos-life.html?_r=0

Rundle, M. (2015, May 8). Facebook's political "echo chamber" is your fault, not theirs. *Wired*. Retrieved from http://www.wired.co.uk/news/archive/2015-05/08/facebook-echo-chamber-study

University of Oregon Libraries. (n d). *Fake news and information literacy*. Retrieved from: https://researchguides.uoregon.edu/fakenews/sift.

Yewah, E. (1993). Ideology and the de/naturalization of meaning in the Cameroonian novel. *Afrika Focus*, 9(3–4).

CHAPTER 7

A Social Experiment: Digital Tech at Its Best & Worst

Figure 7.1. More time with devices = less engagement with the physical world. Pearls Before Swine © 2019 Stephan Pastis. Reprinted by permission of Andrews McMeel Syndication. All rights reserved.

Takeaway: Social media increasingly control the way we engage with each other and the options we have to seek change in our communities.

Our interactions with digital media have provided us with forces pushing in opposite directions. On the one hand, we find ourselves spending much of our time focused on screens—and not in the company of other people.

On the other, through this technology, we have unprecedented abilities to interact with others and use these abilities to build organizations of people who share our perspectives, our experiences, and in some cases our causes.

When these causes offer us the opportunity to act in some fashion with a group we might otherwise not even know about, it gives us a way to think about how power works when you have access to massive, open communication.

It's always helpful to think about these efforts on the continuum of things we should do, things we can do, and things we want to do. Sometimes, when we find that our desires align with those of others, we can work together to achieve change. One of the ironies of the online world is that it is a private sphere, but it thrives on the connections that it makes possible.[1]

But at the same time, we can consider how digital interaction around social causes can lead to some negative social consequences. This can range from not really understanding just what one is reading online, to the rapid reproduction of falsehoods and the experience of instability, to the efforts of bad actors to intentionally flood the digital media space with false claims and highly charged words and images that feed our fears.[2]

In the same way we can find common cause aligned with others with similar interests, the coordination of desires can feed the worst tendencies of people, like the perpetuation of racist, sexist, and homophobic thoughts. Digital media is very good at many things, but not all the things it is very good at are good things.

The Evolution of Social Media: How Did We Get Here?

Where we are today is a product of what came before, and how cultures have always looked for ways to communicate further, better, faster, and to a more satisfying degree. In early internet days there were rudimentary forms of email and networking through sites like CompuServe, which was a privately-owned provider of online services in the 1980s.

It's important to keep in mind the differences between private and public companies. Their legal status and their goals can be driven differently, like the difference between a store that sells books and a public library where you can borrow them and return them.

In the pre-web world, CompuServe set up a networked chat system, with its presence at its greatest peak in the mid-1990s. It also hosted a series of popular online games and message forums on a number of different topics.

There were other private companies that competed for the attention of a mass audience that was waking up to the world of networked computing. CompuServe was founded in 1969, but was eventually joined by Prodigy in 1984 and America Online in 1989. These were mainly services that charged subscription fees, and capitalized on the rapid expansion of personal computers in people's homes.

---

1   Papacharissi (2010).
2   For a great discussion of the relationships between disinformation, misinformation, and malinformation, see Wardle and Derakhshan (2017).

America Online, or AOL as it was known, was famous for its massive use of first floppy disks and then CDs of the program that were sent to millions of home addresses—again, and again, and again. There were so many that people would make art out of them, but the effort was phenomenally successful at building a subscriber base and getting people online.

And much of that base was built around families who were incorporating computers—and the internet—into their home lives for the first time. Once they set up a home modem—essentially a digital phone call to the AOL service—one could participate in an expanding world of digital resources, games, and online discussions. This service also positioned itself as an information resource, a vast collection of journalism, entertainment information, and sports news.

By the late 1990s, they had close to 40 million subscribers. A great number of these subscribers were families, who found the AOL or Prodigy or CompuServe spaces to be educational as well—also supported by the rapid infusion of computer tools into schools. Families felt compelled to not be left behind—an early version of what we now call FOMO (Fear of Missing Out).

In the Interest of "Private Interests"

Marketing and expansion were keys to this developing industry. That meant knowing as much as a company could about their customers, what they were looking for, what—and how much—they would pay for the privilege, and how to turn a profit. They were motivated, in many instances, by basic business principles, the desire to make money and grow, to benefit from and beat the competition.

But understanding customers meant watching them, and seeing what they did, and that meant surveillance. That idea is not new. There have been types of surveillance—where people in power "watched" people who had little—going back through history, in the effort to exert kinds of control. But the incredible pace of technological change meant the ability to surveil—to watch, listen, and collect data—could happen quickly, and on a massive scale.

This is something that, at the time, was not widely understood. To this day, people misunderstand or don't think about privacy, and how the things we do on social networks become data points of valuable information for these private interests.[3]

User-Generated Content—MySpace and Facebook

The early online services created networks of massive, two-way communication platforms. Consumers were also, really for the first time, exclusive producers of the content that populated these platforms— the beginnings of **user-generated content**.

Credited as the first social networking site, Six Degrees started in 1996 with a place to construct an online profile, list school affiliations, and maintain friend lists. They were perhaps too early, because the number of people and the extent of the networks were still too small.

But two other companies—MySpace and Facebook—later began offering similar services, and their timing was a bit better. These two exemplified how the relationship between networked services and users was changing. It was a decided shift away from online companies as content providers, to online companies that engineered ways for users to create the content and interact with each other.

MySpace offered a place and tools for users to develop and maintain their own place on the World Wide Web. Starting in 2003, MySpace established itself with people who wanted to create their own space, including video game developers, music access, and photo storage and sharing. They were also global, and MySpace was the largest social networking site in the world from 2005 till 2009.[4]

---

3   For an eye-opening discussion of privacy and data collection, see Lewis (2017).
4   "The rise and fall of MySpace" by Matthew Garrahan, *Financial Times*.

Its chief users were teenagers and young adults, and they leveraged advertising as the way for the company to make money, rather than users paying for features. Users could embed things like YouTube links into the profiles. As early as 2006, MySpace also started developing ways that their site could be accessed on mobile phones.

MySpace was eclipsed somewhere in the 2008–2009 period by the rapid growth of Facebook. One could argue that MySpace was grounded in the desire of its users to build a profile and communicate about entertainment. Facebook was built on serving a slightly different purpose for their users, from its inception.

Facebook's arguable predecessor was a site called Hot or Not, which started in 2000 and used pictures to elicit discussions and disagreements about the attractiveness of people. Similarly, in 2003, Mark Zuckerberg built a website called Facemash, which he created by literally hacking into the online "face book" student directory photos of Harvard students.

Facemash gave users the opportunity to look at a pair—most frequently women—and choose who was "hotter." After wrangling with the university administration, who shut down the site, Zuckerberg dropped out of Harvard and reinvented the site in 2004 as The Facebook. Originally restricted to Harvard students, it became very popular, and was quickly expanded to other universities.

The growth of Facebook was enormous. It grew to be the most popular social media platform by 2009, which was also the first time that the site turned a profit. By 2011, it was valued at $85 billion.[5]

The notion of exponential growth in digital engagement is an important factor to contend with. Many social media platforms are valuable because of their growth, rather than their ability as a private business to make money. Massive growth—like what Facebook experienced—holds the promise of real future value and staggering profit.

Companies Buying Companies

To better understand social media companies, it helps to look at how they go about merging and acquiring other related media companies.[6] Sometimes that goes well, sometimes not so much.

For example, News Corp[7] bought MySpace in 2005 for $580 Million, winning a bidding war with competitor Viacom Inc., which at the time owned CBS and MTV. But this didn't pan out as well as expected, and the promise of MySpace was eclipsed by the rise of Facebook. News Corp ended up selling MySpace in 2011 for $35 Million—about 5% of the original purchase price.

A year after this sale, in 2012, Facebook bought Instagram. For a billion dollars. For a company that had 13 employees. The motivation for the purchase? According to emails that went public as the acquisition went through antitrust evaluation, Facebook wanted to avoid competition that could hurt its social network. Rather than compete, Facebook acquired.

As Facebook has aged, it's also gained a few warts—complications from privacy breaches and corporate missteps that have drawn public and government scrutiny. For many users, the accumulation of those things triggered a desire to move away from Facebook, to delete accounts or seek social network contact elsewhere, like Instagram. Given that they're both owned by Meta, that's sort of like running away from home out the back door only to go back in the front door.

Among the reasons people gave for departing Facebook was its sheer size. While collecting large numbers of "friends" was an early fad, it also meant that people invited some into their online world

5   Keen (2012).
6   Chapter 3 includes short lists of notable acquisitions by Alphabet (Google), Amazon, Apple, Meta (Facebook), and Microsoft.
7   This conglomerate media company led by Rupert Murdoch owns Fox News, among other media outlets in the U.S., Great Britain and Australia. It also owns *The Wall Street Journal*, Realtor.com, and HarperCollins. Notably, its value is a fraction of any of the companies listed above.

that they wouldn't welcome into their physical lives. It was hard to say no to friend requests and harder to unfriend those you accepted.

According to **Dunbar's Number**, which was proposed by British anthropologist Robin Dunbar, humans can comfortably maintain 150 stable relationships.[8] So if your social network gets above 150 people, it may start to feel unstable and untrustworthy.

Is this true in your experience? Does a large number of people in a social media space become unmanageable, unstable, and unattractive? Since the revenue for these sites is tied to how much time we spend on them, large size may become a disincentive for engagement.

Other Social Media Competitors—and Twitter

Social media platforms have continued to evolve, eliciting a variety of explanations for their success or failure, but almost always having to do with how they fit into the ecology of the social media world.

Take Twitter, for example. Launched in 2006, it was built on the idea of microblogging, the exchange of tweets—short messages of no more than 140 characters initially. The experience is like reading the headlines without reading the stories that they lead.

But how much is it actually used? According to Pew Research Center,[9] only about one in five adults say they use Twitter,[10] and a minority of Twitter users produce the vast majority of tweets—the top 25% of users produce 97% of daily content.

On the up side, Twitter allows users and participants to scan the knowledge environment quickly, see what posted hashtags are trending, and join in threads of conversation to comment on things. On the down side, these conversations made of short bursts of words and images can quickly descend into abusive language, attacks, and reproduced misinformation.[11]

Snapchat, which started primarily as a mobile app in 2011, emphasized images combined with short pieces of text, like a way to share photos with a small group. One pivotal difference was that the user-generated content—snaps—would become inaccessible to viewers in the group after a brief time depending on the user. By 2017, users could change the setting to keep the images available for an unlimited time. They also offered the opportunity of users to develop "stories," to build chronological sequences of images.

The goal of the platform was to encourage a conversational atmosphere. Many social media sites, like Facebook, had built in rating systems (remember Hot or Not?). Users gained gratification from numbers of views and "likes," which made these social platforms transactional.

But what are the effects of these transactions, of having yourself rated based on number of views? There is a great satire of this system in an episode of the series Black Mirror, called "Nosedive" (2016). In this future, everything everyone does all the time is fated on a five-point scale, and people become obsessed with getting higher ratings. And bad things happen.

Like Snapchat and Instagram, the social media site Pinterest dealt mainly in images—pins—that were assemblages of related items in boards. From its start in late 2010, Pinterest also focused on the ability to be used as a mobile app.

And all were supported by advertising.

8   "Dunbar's number: Why we can only maintain 150 relationships" by Tim Smedley, BBC Future.
9   "10 facts about Americans and Twitter" by Meltem Odabas, Pew Research Center.
10   By comparison, another Pew Research Center study found that seven in 10 U.S. adults use Facebook (Gramlich, 2021) and two in five use Instagram (Schaeffer, 2021).
11   For a great exploration of how and why misinformation spreads so swiftly online, listen to the *Reveal* podcast episode, "Viral Lies" (Alcorn, 2022).

The Attention Dilemma—TikTok's Aggressive Algorithm

Here's the problem; you want something in the digital world, you learn about a location that might have what you want, so you go there, consume, and your desire is satisfied, right?

Well, no. Because so many of these sites are built on your continued attention as a measure of success. That means it has to *seem* to give you what you want, but not do it in a way that is completely satisfying. You want more. So you stay on the site, and keep watching.

This is the dilemma faced by the designers of sites like TikTok. You want to give people what they want, but not to the extent that they then move on. Your success is keeping them looking, searching, tumbling down the rabbit hole.

How do these sites do this? They use the details of their computer algorithms to sort and search vast amounts of personal and general information for patterns that indicate what will keep you watching.[12]

Think about your own online habits. How might you give the algorithms useful information about our viewing habits? Do we watch videos all the way through? Do we then give any indication of support, through liking or sharing?

Most importantly, how does the algorithm give us *some* of what we are looking for, but not *everything*—that way, we keep watching, barely noticing how the seconds turn to minutes turn to hours. We become, in some sense, addicted to the experience.

There's also been a significant recent shift between social media where we go for entertainment—to watch but not produce—and others where we might participate. Sites like TikTok can benefit either way we engage. Unlike YouTube, where many users are watching often and contributing rarely, TikTok works to make participation fairly easy. It offers creators background music or audio tracks, and encourages the development of memes that can be imitated and innovated.

By the way, this book's purpose is not to send you running away from social media, but to critically consider what the sites encourage you to do, how they collect information from you to keep you watching, and how they are able to make money from your participation.[13] TikTok, for example, makes money through selling viewers "coins" that they can use to pay creators. Creators can then cash these coins in, with TikTok taking a percentage. And then there are ads as well. TikTok giveth, and TikTok taketh away.

In 2021–22, ByteDance, TikTok's creator and owner, raked in $1 billion of player spending across its various game platforms, and TikTok earned about $4 billion in advertising revenue. In 2022, TikTok announced plans to distribute $200 million to its creators. While that might sound like a lot, it's far less than YouTube's Partner program, which distributed 55% of its revenues—$10 billion—in 2020.[14]

And TikTok is a bit of an outlier from other social media popular in the U.S. in that it's the only platform owned by a foreign company—and government. At this writing, TikTok is owned by a Chinese internet gaming company called ByteDance, with the Chinese government owning a 1% share and claiming one of three seats on the company's board of directors.

It's important to consider what it means for a social media company to be owned by another country—or what that ownership means in relation to politics. Who or what should profit from the manipulation of people's attention spans?

---

12   Algorithms, as Cathy O'Neil has argues, can be judged on two aspects: the quality of their data sets, and their definition of success. See O'Neil (2017) or her Ted Talk on this subject, O'Neil (September 2017).

13   Worb (2022).

14   "Hank Green has a problem with TikTok's creator fund" by James Hale, TubeFilter.

Connect With Me—LinkedIn

Many of the sites discussed above were built around the personal lives of users, creating, sharing, and viewing in a sphere of one's own interests. But the possibilities of using these social structures for more career and professional networking outcomes were realized by LinkedIn. It was launched in 2003, and allowed users to create profiles designed to resemble real world professional relationships.

In 2016 LinkedIn was acquired by Microsoft, with the apparent intention of solidifying connections between the professional profiles and Microsoft's immensely popular business-oriented software suite. LinkedIn also went on its own buying spree of subsidiary businesses, from web applications to eLearning and marketing tools, to coordinate with its core business of professional profiling.

The Evolution Proceeds

Social media platforms are ubiquitous across phones, tablets, and computers. The companies build complex sets of tools to make their social connections both more flexible and more predictable.

But the social media world suffers from a lack of transparency, and a public face that seems both nervously changeable and often unresponsive to criticism. Most of these companies responded to growth by going "public," meaning that ownership is effectively divided into shares that are traded in the stock market. This, again, has significance for how companies respond to both popularity and criticism—and their capacity for controlling our attention and directing our interests.

What does it mean when so much of our interaction happens among online social media platforms? How does that affect our ideas about relating to each other, deciding who is "us" and who is "them"? And how do we negotiate what we do in our own self-interest in relation to the interests of others? How do we think about what social media enables for the common good?

The Common Good—Acting Alone, Acting Together

The drawing together of individuals around a cause is known as a **social movement.** Sometimes a movement gains momentum and sometimes it fizzles. We would not be the first to suggest that inertia—the resistance to change in our speed or direction—poses a challenge to social movements.

But there are other conditions that play into social movements, as well. The better we understand these conditions, the better our chances of finding ways to act in groups that are fulfilling and successful. We also need to come to terms with what the digital world means for how social movements communicate, organize, share information and commentary, and reach out to others to participate.

If We're Acting Alone, Are We Actually Helping?

Coming to terms with movements should be built on coming to terms with how we see ourselves as individuals and as members of groups. So much of culture is grounded in the attempt to free the individual to pursue her or his own path to happiness and fulfillment. But we need to be careful about putting too much emphasis on ourselves as individuals.

Consider what the term "individualism" implies. It would seem to suggest freedom, self-determination, and an ability not to be ruled from outside of your own self.

But there is a more complex way of thinking about individualism that is relevant here: as something that ultimately has roots in the desire to protect self above society.

Ironically, the era of globalization has not led many people to realize their solidarity with others around the world in similar straits. Sometimes it has meant investment in the self at the expense of the group.

The idea here is not to reduce the importance of the individual but to keep its place relative to the groups of which we are members and from which we can draw social and political strength—even when that connection is primarily online. This is one of the reasons that some people question the use of the term "mass" as in "mass communication." It seems in many ways to get in the way of understanding how individuals and groups operate in relation to and through media.

As we move from an idea of individualism to the complexity of our group memberships and global citizenship, the potential is there to find all sorts of different ways of organizing and connecting ourselves with others socially and culturally. And, of course, we need to remember that organizations that seek to build a base online are also impacted by the way algorithms work to either encourage or stymie interactions. The two must function in tandem.[15]

Quite a bit is at stake when we want to start thinking about acting collectively. It's not uncommon for external power to want to keep people separated as individuals, rather than realizing what sorts of ideas they have in common. The existence of the digital world throws a challenge to this.

Maybe.

This is because, quite often, we find ourselves alone sitting in front of a screen. So while we move forward in talking about the potential ability of groups connecting on the Internet to advance a movement consciousness, we need to think through the difference between the person sitting right next to you and the person on the other side of the Internet connection.

Movements and Democracy

Generally speaking, participants in movements are collectively trying to accomplish some kind of change that may be outside of the institutionalized system. Movements seek to accomplish change through moving together, creating symbols of unrest and dissatisfaction, and seeking an accommodation to respond to their collective goals.

Social movements may produce or be inspired by particular leaders (Gandhi, Martin Luther King, Jr., Lech Walesa, Huda Sha'arawi, Gloria Steinem, César Chávez, Nelson Mandela, Greta Thunberg, Emma González), but it is important to remember that *movements* are executed by groups of people who act out of a common interest.

Leadership may be more or less important in any particular movement, but the contemporary digital media era is intensely celebrity focused, and the stories of large groups are harder to tell than the stories of particular individuals.

It is important to see movements not as followers but as collectives—groups of people working together toward change. They are usually people who have very little power on their own, but through the realization of collective desires, and the desire to exert collective power, social movements are able to accomplish quite a bit.

Your experiences and knowledge of history become important components of how you see movements. You may or may not be aware of the Civil Rights Movement, the antiwar movement, or Occupy Wall Street. Or you might have observed something much more local, like a labor strike, a migrant workers' rights movement, or an antipoverty movement.

The Tradition of Social Movements

Social movements have a rich history in Western culture. Consider late 19th- and early 20th-century Progressivism. This movement assembled itself on the idea of "progress" through knowledge that started in the European enlightenment. Progress meant that the application of knowledge would lead to a better standard of living and a better life for all.

Despite the advances in knowledge and technological efficiency, progress seemed to be hard to find for the vast majority of people of little-to-no means. The class of the wealthy and powerful just seemed

15   "You and the algorithm: It takes two to tango" by Nick Clegg, Medium.

to be getting wealthier and more powerful. At the same time, struggle was an enduring reality for the less well off.

Enter the idea of a Progressive social movement. As a collective, this less well-off group sought political and economic reforms to challenge the increasing power of privileged elites and the excessive dominance of business interests.

How did they communicate these concerns? One source was the investigative journalists known as the **muckrakers.** Chiefly through magazine stories, these reporters focused their attention on social ills and high-level corruption.

You have to keep in mind that the media environment was different from what it is today—more responsive to audiences, less consolidated—more diverse really in terms of its ability to reach an audience of like-minded people concerned with social inequities, women's suffrage, child labor, chronic poverty, and even environmentalism.

The social movements that developed around these Progressive notions often involved **collective action**—protests, demonstrations, marches—that were organized through meetings and local contacts. These demonstrations of solidarity were built around numbers, but they were also built through organizers. These were planned actions.

The New Social Movements

Social movement theorists started calling the movements that appeared in the 1960s the "new social movements." These were characterized by a shift in focus from the material inequality issues of the Progressive Era movements to a more specific focus on identity and social justice. They sought an orientation that was more about the issues and less constructed in terms of the political system.

Where the Progressives actually launched the most successful third party in American history, the new social movements of the 1960s resisted affiliation with a political party (often because of a deep skepticism about the system of political power in general).

And where the Progressives used magazines and newspapers as their main vehicle of communication, the position of the new social movement actors grew out of much higher levels of advanced education engagement with mass media—chiefly television. Despite the increasing corporatization of the mass media, the leaders of these movements in many cases understood how the media could be "played."

TV news at this stage was considered an obligation of the networks and operated as an expense rather than just another source of profit as we see in the contemporary TV news environment. And the movements knew that social unrest and spectacle made for great network news TV.

From Collective Action to Connective Action

Relative to the media, these movements have migrated from the theater of network television, through the development of the 24-hour news cycle on cable TV news, to the interactivity of the World Wide Web, and, finally to the viral influence afforded by social media like Twitter. That notion of interactivity is a significant part of the way communication among people involved in social movements—and recruitment of new followers—has changed.

There were some alternative media in the new social movement era, but they were underfunded and in the shadow of corporate media, so the logic was to become an object of interest to the corporate media and get attention to the causes as a result. The digital era changed that, since the communication platform allowed for massive, complex networks of interactivity.

Paying attention to the digital divide tells us that the means of communication are always an economic privilege, but the scale has shifted quite a bit in terms of the resources needed to become a voice or, for the interest of movements, a chorus of voices.

Movements have changed as a result of the interactive digital network.[16] Perhaps most significantly, the logic of collective action has been replaced by the logic of ***connective* action.**

Where the *collective* included individuals as a mass of anonymous participants, *connective* logic allows for a more personal participatory experience. It is less dictated by, say, the relationship between a TV network and its story subjects, and more a product of interactivity and inclusive individuality.

And where collective action today may begin with physical meetings of groups at the local level and move to online organization, connective action finds its strength in online connections that may also organize physical world protests.

Historically, collective action has led to events such as the 1963 March on Washington, which drew more than 250,000 civil rights protestors to the nation's capital. The Civil Rights Movement was championed by the charismatic Martin Luther King, Jr., who became the face for the movement.

The 2017 Women's March in Washington drew over 470,000 people and was streamed live on YouTube, Twitter, and Facebook, in addition to broadcast and cable television coverage. And, building through online connections, participation totaled between 3 and 5 million marchers at rallies across the globe.

By comparison, the 2014 Black Lives Matter march on Washington drew only about 25,000 protestors, but it was buoyed by the deployment of the #BlackLivesMatter hashtag by millions of social media users. But deployment of such hashtags also comes with a warning: Their overuse tends to drown out the voices of those whose most need to be heard. They also can be hijacked by groups who oppose the movement.

Hashtag activism, as this has become known, is the idea of social media being used to raise awareness of a wide range of causes, simply by offering their support for, or opposition to, the movement in posts on social media.

Critics of hashtag campaigns have raised concern that the individual act of liking a hashtag, sharing it, or using it in an original post does nothing more than briefly raise visibility of an issue. The concern is that hashtag campaigns make users think they are doing something, but nothing is ever accomplished and attention to the issue wanes when our short attention spans are directed elsewhere.

Proponents of hashtag campaigns praise their ability to allow anyone to participate to whatever degree they choose—from just reposting the hashtag to organizing physical-world protests—as well as their capacity to quickly elevate an issue at a global level.

There will likely be some sort of new digitally constructed social movement that will pop up while you are reading this book. Look carefully at how it negotiates its identity. These movements often draw attention through coverage by legacy media but at the same time exert power and identity through social and other digital media.

Are We "Anonymous"?

Finally, a kind of reflective commentary on the nature of social movements has arisen through a series of actions that are grounded in the intersection of digital culture, mass media, and an ethos that combines subversive "hacking" with the more traditional idea of exposing social wrongs.

Global digital culture in general is now being faced with the consequences of this idea. The interconnectedness of the Internet requires a substantial investment in digital security schemes. We have all become so dependent on digital infrastructure that we all become vulnerable if and when the net is hacked.

But hacking—and the "leaks" that result—sometimes shines a light on covert activities, some of which are covert because their existence *could* lead to more massive movement activity.

16   Bennett and Segerberg (2013).

One could argue that this interconnected web of hackers and digital distributors has taken up the role that was once occupied by the muckraking magazines of the Progressive Era and the oppositional mainstream media of the new social movement era.

Once upon a time (OK, it was 1971), *The New York Times* published a controversial set of documents called the Pentagon Papers, which revealed the military account of what was happening in the Vietnam War. This military description differed considerably from what was actually happening. The United States had bombed two other countries—Laos and Cambodia—without letting the U.S. public know about it. This *New York Times* publication was not itself a movement, but it led to a great expansion of the antiwar movement in the United States.

We move forward a few decades and the relative position of *The New York Times* has changed. In 2002, it published prominent, front page stories about weapons of mass destruction in the possession of Saddam Hussein, who at the time was leader of Iraq. The stories were fundamentally used to support the U.S. invasion of Iraq in spring 2003—despite their lack of validity. *The New York Times,* in this case, was not leading the fight against propaganda, as with publication of the Pentagon Papers, but it had been co-opted as part of the government's propaganda machine.

With mainstream media no longer leading the way, a series of investigative, internet-connected organizations have learned to use hacking (or leakers/whistle-blowers) to locate buried evidence of covert activity and bring it to light.

For example, in the spring of 2010 WikiLeaks released footage of a July 2007 airstrike where an American helicopter shoots several people in Iraq, including several journalists. This video was made widely available through the WikiLeaks website and quickly spread through the network of video torrents and other person-to-person file exchanges.

This brings us to a significant movement worth additional consideration because of the way it changes the notion of participants.

Anonymous is a loose organization of **hacktivists** (hacker-activists), a decentralized group of participants who have taken the idea of identity in a movement and made it into a politics of anonymity with its own agenda.

Remember that the early movements had leaders and hierarchies and that the new social movements tried to focus on numbers of members rather than leaders? This was true even though mainstream media would still focus on leaders, due to the difficulty of covering truly mass movements.

Anonymous makes this even more difficult, since the image chiefly associated with it is a Guy Fawkes mask (look him up!). But its more direct source was the graphic novel and film *V for Vendetta*. In this story, an anonymous character in this mask leads a series of increasingly violent protests against unchecked governmental power and manipulation.

Anonymous capitalizes on the World Wide Web's ability to make identity into an optional object, as this group of borderline illegal hacktivists creates stunts around shutting down web locations via Distributed Denial of Service attacks.[17] Its chief targets are any organization—governmental or private—that is advancing censorship. Anonymous often uses the tagline, "We are Anonymous. We are Legion. We do not forgive. We do not forget. Expect us."

So the group's power as a movement is grounded in our not knowing who the members are specifically or even really knowing what philosophy explains their actions. Their actions use the power of the Internet to advance their causes in a way only the complex digital network can support.

17 DDOS attacks, as they are known, are a relatively complicated, concerted effort in which services offered by an online system are disrupted by two or more sources tapping into and effectively blocking access to it. For more information and current examples of DDOS attacks, look up the term in a search engine.

Are they acting in your best interests? Are they a kind of digital general strike, protesting for individual determination in the face of state and corporate power? Are they agents of vigilantism, threatening to undermine the Internet, like a community watch patrol gone wild?

The value of Anonymous hacktivists, good or bad, may be determined by their next action. Like other kinds of movement politics, you may find yourself in sympathy with their causes—or you might find yourself contemptuous of their self-righteousness and pointlessly disruptive behavior.[18]

The Power of Power

In your experience, you may have sensed the power of a large group, perhaps organized, perhaps less organized than suits your preferences. Your experiences with movements may change depending on your own participation.

Being moved to act with a group of like-minded people can be exhilarating, and it can lead to great accomplishments in the face of regimes of power. Being a good student of history will allow you to see how movements in the past have accomplished great things, often making living conditions better for large groups of previously disempowered people.

In that history you will notice the critical role that communication plays in accomplishing a movement's goals. Movements are still learning what is made possible by the shift from collective action to connective action. Inequities in power are inevitable, so there will be more to come.

Summary

Social media is often heralded as a bastion of freedom, allowing oppressed peoples to share their experiences with the world in an instant. But that promise is clouded by the ever-increasing monetization of our thoughts and ideas by the companies that dominate digital advertising. Instead of expanding our minds, social media too often allows us to narrow them by choosing who and what we follow, figuratively feeding us back to ourselves. The result sometimes is collective action that leads to change, but more often it's inaction as we are continually redirected back to our own needs, wants, and beliefs.

Key concepts

Anonymous
Collective action
Connective action
Dunbar's Number
Hacktivists
Hashtag activism
Social movements
User-generated Content

References

Alcorn, S. (2022, January 1). *Viral lies*. Reveal. Retrieved from: https://revealnews.org/podcast/viral-lies-2022/

Bennett, W. L. & Segerberg, A. (2013). *The logic of connective action: Digital media and the personalization of contentious politics*. New York, NY: Cambridge University Press.

Clegg, N. (2021, March 31). *You and the algorithm: It takes two to tango*. Retrieved from: https://nickclegg.medium.com/you-and-the-algorithm-it-takes-two-to-tango-7722b19aa1c2

Garrahan, M. (2009). The rise and fall of MySpace. *Financial Times*. Retrieved from https://www.ft.com/content/fd9ffd9c-dee5-11de-adff-00144feab49a

---

18   For a great way to think about many of these issues, in addition to the movie V For Vendetta, check out Mr. Robot—search for it online.

Gramlich, J. (2021, June 1). 10 facts about Americans and Facebook. *Pew Research Center*. Retrieved from https://www.pewr esearch.org/fact-tank/2021/06/01/facts-about-americans-and-facebook

Hale, J. (2022, January 4). Hank Green has a problem with TikTok's creator fund. *TubeFilter*. Retrieved from https://www. tubefilter.com/2022/01/24/hank-green-tiktok-creator-fund-earnings-per-view/

Keen, A. (2012). *Digital vertigo: How today's online social revolution is dividing, diminishing, and disorienting us.* New York, NY: St. Martin's Press.

Lewis, R. (2017). *Under surveillance: Being watched in modern America.* Austin, TX: University of Texas Press.

Odabas, M. (2022, May 5). 10 facts about Americans and Twitter. *Pew Research Center*. Retrieved from https://www.pewresea rch.org/fact-tank/2022/05/05/10-facts-about-americans-and-twitter/

O'Neil, C. (2017). *Weapons of math destruction: How big data increases inequality and threatens democracy.* New York, NY: Broadway Books.

O'Neil, C. (2017, September). "The era of blind faith in big data must end" (video). *TED Conferences*. Retrieved from https:// www.ted.com/talks/cathy_o_neil_the_era_of_blind_faith_in_big_data_must_end?language=en

Papacharissi, Z. (2010). *A private sphere: Democracy in a digital age.* Maiden, MA: Polity.

Schaeffer, K. (2021, October 7). 7 facts about Americans and Instagram. *Pew Research Center*. Retrieved from https://www. pewresearch.org/fact-tank/2021/10/07/7-facts-about-americans-and-instagram/

Smedley, T. (2019, October 1). Dunbar's number: Why we can only maintain 150 relationships. *BBC Future*. Retrieved from https://www.bbc.com/future/article/20191001-dunbars-number-why-we-can-only-maintain-150-relationships

Wardle, C. & Derakhshan, H. (2017, October). *Information disorder: Toward an interdisciplinary framework for research and policymaking.* Council of Europe. Retrieved from: https://rm.coe.int/information-disorder-toward-an-interdisciplinary-framework-for-researc/168076277c

Worb, J. (2022). How does the TikTok algorithm work? *LaterBlog*. Retrieved from: https://later.com/blog/tiktok-algorithm/

CHAPTER 8

Much to Lose: The Symbiotic Relationship among Journalism, Technology, & Democracy

Figure 8.1. Can a democracy survive without journalists to interpret, document, and draw the public's attention to the important issues that continually shape and change our world? If no one's paying attention, it doesn't matter—the government and corporations end up with unchecked power. As goes journalism, so goes democracy. Pearls Before Swine © 2022 Stephan Pastis. Reprinted by permission of Andrews McMeel Syndication. All rights reserved.

Takeaway: Only time will tell if democracy can survive a worldview that favors opinion over fact, individual over community, and corporate rulemaking over legislation. We have a role here …

Quick: Who was the first U.S. president to use social media to speak directly to the public? And which president broke a Guinness World Record by amassing over one million Twitter followers in five hours?

You may be surprised to learn that neither was Donald Trump.

Barack Obama was one of the first presidential candidates to capitalize on the power of social media, with his 2008 campaign using a range of social media channels to communicate his ideas and raise money.

Part of that attraction was likely demographic, as Obama, who was 47 at his 2009 inauguration, was one of the youngest presidents in U.S. history.[1] As he entered office, Obama was the first president to do so with a smartphone in his grasp. He made it clear that he would continue to use his beloved BlackBerry,[2] which he favored over the iPhone for his personal use.

As president, Obama had no great love for the press, which he termed "the filter."[3] There was a point during Obama's second term that photographers were prohibited from taking their own pictures at events. Instead, only one photographer was allowed to photograph the president in the White House and only those approved photos were distributed for use by the media.

Obama's standoffish approach to the media was nothing new; many presidents had successfully controlled the flow of news out of the White House. It was particularly easy when there were only newspapers, radio, and three television networks in the picture. The president could have roughly the same message hit everyone.

Digital media cut out traditional gatekeepers, however. Anyone could comment and participate. And social media users could choose whose messages they wanted to see in their feeds.

Rather than relying on the White House Press Corps to interpret and disseminate the news, the Obama Administration took to social media pathways to address the public directly. The @POTUS handle[4] was added to Twitter May 18, 2015, hitting one million followers within its first five hours (the aforementioned Guinness World Record). President Obama's first tweet:

Hello, Twitter! It's Barack. Really! Six years in, they're finally giving me my own account.

That was followed by a quick welcome from @WhiteHouse, the up-until-then Twitter account for the nation's executive branch:

President Obama. In the Oval Office. Tweeting.

#WelcomeToTwitter, @POTUS!

---

1   Teddy Roosevelt was the nation's youngest president. He was 42 at his inauguration. John F. Kennedy was 43, Bill Clinton and Ulysses S. Grant were each 46.
2   The BlackBerry, launched in 1999, was also known as a personal digital assistant. It evolved into a hand-held computer with a small screen and a full QWERTY keyboard, which allowed users to make phone calls, surf the web, and send and receive email. Obama's version was the BlackBerry 8830 World Edition, which launched in May 2007, just over a month before the first iPhone hit the U.S. market.
3   "Obama's media machine: State run media 2.0?" by Devin Dwyer, for ABC News.
4   That's President of the United States, in case you've ever wondered.

In theory, that democratized presidential communication by allowing the president's perspectives to reach a targeted audience without interpretation, added context, or mediation by journalists.[5] Yet this was significant for three not-so-good reasons:

1. It was exclusive. Rather than potentially reaching the entire nation via traditional media channels, use of social media limited the message reach to those who had the knowledge and ability to engage with it.[6]
2. The end-run around journalists, whose job is to provide context to the news, meant less accountability for the administration, and fewer readers and listeners tuning in to absorb and consider that criticism.
3. It was preaching to the choir. That old saying means that you're talking to people who already share your beliefs. By pre-filtering "news" and using social media channels instead of the press, it's likely the Obama Administration was largely hitting its supporters. And those who opposed the Obama Administration and took to social media channels themselves were only hitting their supporters.

In short, the exclusive, divisive nature of social media had the potential to fragment the U.S. public and derail democracy. One journalist framed the shift that took place during the Obama Administration as follows:

> Even before Obama and his new politics burst onto the scene in 2008, we knew we lived in a country that is evenly but drastically split in its worldview. Now that the opposition can rewrite the news as it's happening, the two sides can essentially live in separate realities—which is perhaps what we've always wanted. It's becoming clear, though, that two casualties of this new order are efficacious, fact-based governing and an independent, fact-based press.[7]

This is the expected outcome of a magnifying-glass-and-ant approach to news dissemination, one that allows people to choose what information they want to consume and from whom. It's also the result of our fascination with the cult of personality, which results in individuals gaining larger, louder channels that effectively drown out the voices of journalists— the very people whose constitutionally-protected work helps protect and maintain the historical record.

The effects of "social" media on the U.S. democracy started to become painfully evident during the 2016 presidential election, roughly matching the pace with which the Frightful Five cemented their place in our lives and the global economy.

The Evolution of Presidential Communication

Fast-forward to 2016, and social media ended up playing a significant role in the election of President Donald Trump and largely defined his entire presidency.

At 70, Trump was the oldest president in U.S. history at his inauguration.[8] He was, however, unlike any other president in his intensive use of social media to directly speak to the world.

5   "Here's how the first president of the social media age has chosen to connect with Americans," by Juliet Eilperin, *The Washington Post*.
6   Demographic factors that affect technology engagement in the U.S. are covered in depth in Chapter 12, "Haves, Almost Haves, and Have Nots: The Domestic Digital Divide."
7   "The presidency and the press," by Reid Cherlin, *Rolling Stone*.
8   Joe Biden was 78 at his inauguration, eclipsing Trump's brief hold on that record. Prior to Trump, Ronald Reagan had been the oldest present at his 1981 inauguration at the age of 69, breaking a record held 140 years by William Henry Harrison.

Primarily, that was because of the source of the tweets. While Obama's tweets were drafted with the care of a public address, vetted, and very unlikely to have come directly from his fingertips, Trump was an avid Twitter user who had tweeted tens of thousands of times since joining the platform as @ RealDonaldTrump in 2009.

By the time Obama left office in January 2017, @POTUS had amassed more than 13 million followers. That account—*followers intact*—was transferred to Donald Trump when he took office in 2017.

While Trump used @POTUS early in his presidency, he swiftly shifted to primary use of his pre-existing personal account, @RealDonaldTrump. From there, as president, Trump tweeted and retweeted more than 25,000 times. He adopted a conversational, sometimes adversarial tone, gaining ground as a master of quips, near-comical slips, and nicknames that stuck. In unprecedented unpresidential style, he fired people via Twitter, chastised other nations, blocked users who offended him, and deleted his own tweets in defiance of the historical record.[9] As he tweeted midway through his first year of office, "My use of social media is not Presidential—it's MODERN DAY PRESIDENTIAL."

Trump also was known for using provocative tweets as a technique to distract journalists, who inevitably would turn their attention away from other news to report on particularly inflammatory comments.

It was not unusual for Trump to send dozens of tweets each day, setting a record of 142 tweets in one day as he weighed in on the proceedings of his first impeachment trial. In mid-2020, while up for re-election and in the midst of civil unrest relating to the death of George Floyd, Trump broke that record, sending 200 tweets and retweets in one day.[10]

@RealDonaldTrump ultimately had nearly 89 million followers, placing it in Twitter's top 10 highest-follower accounts.[11]

Trump's love/hate affair with Big Tech ended abruptly in 2021, when he was unceremoniously suspended, banned, and/or entirely removed from most major social media platforms.

Interesting. But What Does This Have to Do With Democracy?

This is a good point to back up and review some **tenets of democracy**:

- Free and fair elections

- Accepting election results

- Peaceful transfer of power

- Rule of law

- Tolerance, cooperation, and compromise

- Defense of the Constitution against all enemies, foreign and domestic

There is growing evidence that social media has had a significant and evolving role in eroding each of those foundations.

---

9   "As President Trump tweets and deletes, the historical record takes shape," by Rachel Treisman, NPR.
10 "Trump broke his all-time tweeting record amid nationwide protests," by Connor Perrett, *Insider.*
11 While impressive, Trump's follower count was well below the top account of @BarackObama, which at this writing had 132 million followers. Find this info on Wikipedia under "List of most-followed Twitter accounts."

Donald Trump did not win the popular vote in 2016, but he did win the majority of state-allocated electors, in what's known as the Electoral College. With that evident, Hillary Clinton conceded the election, honoring more than two centuries of precedence. President Obama welcomed Trump to the White House on inauguration day, embodying the peaceful transfer of power.

Trump's election ultimately was found to have been facilitated by an aggressive, sustained social media campaign conducted by Russian agents. The aim was to steer voters away from Clinton by whatever means possible. An independent, international news outlet summarized the impact across platforms:

> Between January 2015 and August 2017, Facebook linked 80,000 publications to the Russian company Internet Research Agency through more than 470 different accounts. At the same time, a total of 50,258 Twitter accounts were linked to Russian bots—fake accounts programmed to share false information—during the 2016 election period. The bots are responsible for more than 3.8 million tweets, about 19% of the total tweets related to the 2016 US presidential election. Approximately 80% of these bots behaved in a way that supported Donald Trump, mostly using the hashtags #donaldtrump, #trump2016, #neverhillary and #trumppence16.[12]

Subsequent investigation of that interference resulted in the indictment of 12 Russian military intelligence officers, as well as the indictment and conviction of Trump's 2016 campaign chair, Paul Manafort, and his administration's national security advisor, Michael Flynn.[13]

Beyond that, journalists from several international newspapers worked together to unspool accusations that a data firm, Cambridge Analytica, mined Facebook to build psychological profiles of voters. According to a *The New York Times*—and confirmed by Facebook itself—the consulting firm acquired data from nearly 90 million Facebook users, most of whom were in the United States. That data, in turn, was used to identify and reach particular American voters via social media with a targeted spread of misinformation.[14]

Blowback on Facebook was swift, with its top security official leaving and some users deleting their accounts. That didn't last long. Mostly there was a collective shrug, with Facebook users being resigned to the amount of personal data "free" platforms collect. One person targeted in the breach told a reporter, "I've come to grips with the fact that you are the product on the internet. If you sign up for anything and it isn't immediately obvious how they're making money, they're making money off of you."

Another, on being told that Cambridge Analytica had collected personal data from her account and those of all her friends and family, said it was unlikely she'd be leaving Facebook as a result: "I'm just too nosy to stay off it."[15]

Whose Speech Is Protected by Social Media?

While the social media platforms were not specifically implicated in shaping the outcome of the 2020 presidential election, their very nature had resulted in a bitterly divided country—some passionate followers of Trump determined for him to stay in office and others passionate in their desire to get him out.

Trump again lost the popular vote, but this time he also lost the Electoral College, and Joe Biden was declared president-elect after four tense days of vote counting and recounting. Despite lack of

---

12 "Fact check US: What is the impact of Russian interference in the US presidential election?" by Sophie Marineau, *The Conversation*.
13 Both of whom were pardoned by Trump in his last month as president.
14 "Cambridge Analytica and Facebook: The scandal and the fallout so far," by Nicholas Confessore, *The New York Times*.
15 " 'You are the product': Targeted by Cambridge Analytica on Facebook," by Matthew Rosenberg and Gabriel J.X. Dance, *The New York Times*.

proof of anything but a free and fair election, Trump never conceded the election and used his social media platforms, notably Twitter, to amplify claims that he was the victor and that the win had been stolen by "voter fraud."

Rather than championing a free and fair election, accepting election results, and committing to a peaceful transfer of power, Trump rallied his supporters repeatedly via social media. In December of 2020, he encouraged them to join a "Stop the Steal" rally in Washington, D.C., on the day Congress would be certifying the election. Many of the president's supporters organized and arrived that day due to what they perceived as direct invitation from the president, *delivered via tweet*:

> Statistically impossible to have lost the 2020 Election. Big protest in D.C. on January 6th. Be there, will be wild!

That rally drew thousands of supporters to Washington, resulting in multiple breaches of the Capitol building, hundreds of arrests, countless injuries, and five deaths.

This gets us to the very active role social media platforms took after a near-constitutional crisis unfolded live across the globe via traditional and social media channels. With Trump's posts indicating a lack of interest in interrupting the potential for further mayhem, social media companies took action. The result was **deplatforming**, removal of a person's access to social media channels and thus ability to reach other users. In this case, that person was the president of the United States, who within days was suspended or permanently banned from Twitter, Facebook, Instagram, and YouTube.

After removing @RealDonaldTrump and banning Trump from tweeting to his nearly 89 million followers, Twitter began a lively game of whack-a-mole as the then-president shifted from account to account to speak directly to the platform's users. Trump quickly turned to @POTUS, which had amassed nearly 33 million followers, tweeting "We will not be SILENCED!" and "Twitter is not about FREE SPEECH."[16]

He also used his campaign account, @TeamTrump, which was also promptly booted by Twitter. Before removal, that account directed its 2.3 million followers to Parler, an independent social media platform launched in 2018 that largely attracted conservative users.

At that point, Apple and Alphabet opted to deplatform Parler from their mobile app stores, and, on January 10, 2021, Parler disappeared completely when Amazon canceled its hosting service.

It's critical to our understanding of technology's interplay with democracy that these limits to free speech were taken not by the government, but by Twitter, an independent publicly-held company, and *four* of the largest tech companies in the world—Alphabet, Amazon, Apple, and Meta. Other highlights of this unilateral corporate action included:

— Twitter banning accounts of key Trump supporters

— Facebook and Instagram banning the "voice of Trump," preventing him from posting videos / interviews on other people's accounts

— Twitter breaking from policy set in 2017 and removing all @POTUS followers when the Biden administration took over the account January 20, 2021.[17] That effectively set an immediate limit on Biden's early audience as he had to restart the account.

16   "Twitter deletes new Trump tweets on @POTUS, suspends campaign account," by Reuters staff.
17   By the end of his first day in office, Biden had gained nearly a million followers, which grew to nearly 24 million midway through his second year. Details from "Biden takes over POTUS Twitter account, inheriting a blank slate from Trump," by Rachel Lerman, *The Washington Post*.

- YouTube restricting Trump and advisor Rudy Giuliani from profiting from partnership ad revenues, subscriptions, and donations on the platform

- Meta CEO Mark Zuckerberg blaming Trump and users, not the technology in testimony before Congress, saying:

> I think the responsibility lies with the people who took the actions to break the law and do the insurrection. Secondarily, also with the people who spread that content, including the president but others as well, with repeated rhetoric over time, saying that the election was rigged and encouraging people to organize, I think that those people bear the primary responsibility as well.[18]

It's important to note that, prior to January 6, 2021, Trump's provocative speech on Facebook had been tacitly approved via the company's **XCheck** program. This practice exempts the speech of millions of celebrities, star athletes, political leaders, and high-follower influencers from many platform rules. Flagged infractions that would result in the suspension or removal of average users are not sent through Facebook's normal review.

The actions of social media platforms following January 6, 2021, exemplify the nature of digital participation: you have a right to publish online, as long as you play by the rules of your host—community standards defined in user agreements. They also pose an interesting question about who gets to define free speech and whether the social media companies themselves can still be considered to be content-neutral.

Response from leaders of other democracies were quick to criticize not only Trump's overt opposition to conceding a lost election and peaceful transfer of power, but also his rapid eviction from social media.

European Union Commissioner Thierry Breton called the Capitol riot social media's "9/11 moment":

> The unprecedented reactions of online platforms in response to the riots have left us wondering: Why did they fail to prevent the fake news and hate speech leading to the attack on Wednesday in the first place? … The fact that a CEO can pull the plug on POTUS's loudspeaker without any checks and balances is perplexing. It is not only confirmation of the power of these platforms, but it also displays deep weaknesses in the way our society is organized in the digital space.[19]

German Chancellor Angela Merkel called removal of Trump's accounts "problematic," noting intervention in the freedom of expression should be defined by legislators, not by corporate decisions. French Finance Minister Bruno Le Maire agreed that the state, and not "the digital oligarchy," should be responsible for regulations, and called the technology companies a threat to democracy.

European Parliament member Manfred Weber was more direct, pointing to potential action by the European Union: "We cannot leave it to American Big Tech companies to decide how we do and do not discuss, what can and cannot be said in a democratic discourse. We need a stricter regulatory approach."[20]

Following dismissal of a lower court ruling that Trump violated the First Amendment rights of users he blocked on Twitter, Supreme Court Justice Clarence Thomas also accused Big Tech of having vast and unchecked power:

---

18   "Zuckerberg blames Trump, not Facebook, for the Capitol attack," by Taylor Hatmaker, TechCrunch.
19   "Capitol Hill—the 9/11 moment of social media," by Thierry Breton, *Politico*.
20   "Merkel among EU leaders questioning Twitter's Trump ban," by Perre-Paul Bermingham, *Politico* Europe.

As Twitter made clear, the right to cut off speech lies most powerfully in the hands of private digital platforms. The extent to which that power matters for purposes of the First Amendment and the extent to which that power could lawfully be modified raise interesting and important questions. It changes nothing that these platforms are not the sole means for distributing speech or information. … In assessing whether a company exercises substantial market power, what matters is whether the alternatives are comparable. For many of today's digital platforms, nothing is.[21]

Where Is Technology Taking Our Democracy?

Political theorist Benjamin Barber set forth **three scenarios for the future of technology and a strong democracy**:[22] Pangloss, Pandora, and Jeffersonian. Each is accompanied by what he called a "psychological mood":

Pangloss = Complacency
Pandora = Caution
Jeffersonian = Hope

The Pangloss Scenario has its roots in the distant past. Pangloss was a character in the French satire *Candide* by Voltaire, which dates back to 1759. Pangloss was a tutor for Candide, who was sheltered from the world on a French estate. Pangloss painted the world as a utopian paradise.

The Pangloss view of the future is relentlessly upbeat. It's the Meet-the-Robinsons-by-Disney view of the future. Blue skies. People floating around in bubbles. Buildings grown, rather than built. Clean. Bright. Perfect.

In the Pangloss Scenario, technology's impact on physical communities is that it solves problems, particularly communication and economic problems. For example, the adoption of television broke up the radio monopoly, the adoption of cable television broke up the big-three network monopoly, and email and online "bill-pay" broke up the U.S. Postal Service monopoly. And digital media are breaking up the tyranny of experts imposed by legacy media and other traditional institutions. Barber termed this perspective as "cyber-enthusiasm."[23]

That's the upside.

The downside is the inequality of a dominant worldview that favors corporations and a market-driven model. If your problem can't be solved by technology, then it very well may be "glossed over" by a dominant worldview that doesn't accept that technological fixes don't work for everyone and everything.

In the Pangloss view, the business model for the web benefits companies and sites that make money—Alphabet, Amazon, Apple, Meta, Microsoft—not those with a strong social value, like traditional journalism.

The primary concern for democracy is that technology favors those who engage with it and distracts them from helping those who do not, or cannot. Old people? The poor? Minorities? Marginalized. Here's what Barber had to say about the Pangloss view:

> At best, the market will do nothing for uses of the new technology that do not have obvious commercial or entertainment or corporate payoffs, and, at worst, will enhance uses that undermine equality and freedom.[24]

21   "Justice Clarence Thomas takes aim at tech and its power 'to cut off speech'," by Bobby Allyn, NPR.
22   Barber (2006).
23   Barber (1998), op. cit.
24   Barber (2006), op. cit., p. 194.

The Pandora Scenario comes from what's probably a more familiar story from Greek mythology. Pandora was the first woman. She also was given the task of protecting a sealed container, which contained all the evils of the world. She was instructed not to open it. So, what did she do? She opened it, unleashing evil into the world to plague humankind.

In the Pandora view, the internet is a tool that poses a peril to democracy. It is full of humankind's worst intentions and distracts people from the very real needs of their physical world. Barber termed this perspective as "cyber-pessimism."[25]

Part of the issue relates to the nature of the World Wide Web: The technology promotes good and evil without judgment and in relatively unequal doses.

We are blessed with tremendous freedom of information that can facilitate remarkable civic engagement.

We are also cursed with an overwhelming flood of information that sends us scrambling for higher ground—and that higher ground generally focuses us on ourselves. Individualism is not a friend to democracy.

The Pandora view focuses us on the tremendous costs to our privacy that "free" services of the web deliver. Corporations own the primary portals and most popular search engines, and they control the technology that monitors your every move. The government—at least in the United States—has provided little to no restriction as to how your personal information is used.

The greatest threat to democracy in the Pandora view is that governmental and corporate power will be unchecked in a society that favors technologies that cultivate individualism. Read this assessment by Barber:

> There is no tyranny more dangerous than an invisible and benign tyranny, one in which subjects are complicit in their victimization, and in which enslavement is a product of circumstance rather than intention. Technology need not inevitably corrupt democracy, but its potential for benign dominion cannot be ignored.[26]

The Jeffersonian Scenario is anchored in the foundations of democracy. Thomas Jefferson was the primary author of the Declaration of Independence and the third president of the United States.[27]

Jefferson's view was that the inadequacies of democracy are best remedied by more democracy; if people are struggling with their freedoms, give them more freedom

In theory, digital media should offer us far more ways to express ourselves and participate in our communities. While legacy media have always favored active programming and passive spectatorship, digital media tend to favor selective programming and active participation.

In theory.

To fully understand the downside of this equation, we need only think back to the downsides of the Pangloss and Pandora views: inequality, unchecked power, and individualism.

What good does it do to have greater participation if people are only going to be focused on the things that interest them and the beliefs they hold? Individualism tends to anchor those interests and beliefs, not change them.

The risk to democracy is an increased level of **civic incompetence,** where people make their opinions known, informed or not. That said, the Jeffersonian view is that civic incompetence is not a reason to exclude people from participation in a democracy. The role of the government and technology in a

---

25 Barber (1998), opt cit.
26 Barber (2006), op. cit., p. 195.
27 You already knew this … right?

democracy is to foreground important news—amidst the noise and distraction of digital media—and facilitate discussion.

What is Barber's take? The Jeffersonian Scenario, as good as it sounds, is the least likely of the three to take hold. Here's his summation:

> A free society is free only to the degree that its citizens are informed and that communication among them is open and informed … In order to be something more than the government of mass prejudice, democracy must escape the tyranny of opinion.[28]

Barber's skepticism that we could rise above our divisions reflects our existence in a post-narrative world where overarching stories, an informed electorate, and the quest for Truth have been eclipsed by personal opinions, divisive ideologies, and individual truths.

How Do Journalism, Media, & Technology Support Democracy?

Journalists were once part of the defense system that protected the public from information overload, along with social institutions such as law, finance, healthcare, religion, schools, and the family. Each of these institutions has been charged with enforcing particular standards that relate to the meaning and filtration of information. And each has been modified—arguably weakened, if not broken—by our love affair with digital technology.

This is a reasonable point to tease apart the terms media and journalism, which often present as interchangeable. They're not.

Culturally, we have had a long and complicated relationship with the term **media**, a remarkably flexible word that takes on a range of indistinct meanings. It's a plural of the word medium, which at its base means something in the middle. That's the nature of mediated communication—there's something between you and the source.

One common use of the word media refers simply to content, the words, images, music, and such we consume daily via analog technologies, like paper, and digital technologies and platforms, like apps and the World Wide Web.

The **medium** itself—the device or platform conveying a message—is the source of mediation. Television, for example, is a medium that conveys information in one direction. Social media, the blanket term for participatory platforms, is one that facilitates multi-directional conversations involving lots of voices on an infinite number of topics.

The collective term "**the media**" also is used to describe the news and entertainment organizations that create and distribute content for mediated communication. That use focuses more on the influence of ownership on the message, more than the relative neutrality of the medium conveying it. For example, "mainstream media" has become popular in the past few years to refer to the perceived desire of traditional media—large, established news organizations—to maintain society's status quo. It's a slur against media and a coded call for consumers to seek alternative news organizations.

Use of the term media as an intentional slight has the effect of negating the value of media content, which in turn denigrates the work those who produce it for a living. That hyper focus on ownership and technologies hinders our ability to understand the importance of content producers.

So, what the word media doesn't describe is journalism or journalists. At its best, **journalism** is not a profession or a job; it is a tradition, a craft, a calling, and a mission. **Journalists** are not supposed to work for corporations, companies, newspapers, or websites; they're supposed to work for the public. Good journalism should provide the information necessary for people to better understand their communities, their states, their country, and the world.

28   Barber (2006), op. cit., p. 196.

Journalism has traditionally assumed a role relating to the maintenance and evocation of a society's values, particularly with regard to a community's historical record. It was with newspapers that the first glimpses of our democracy appeared, via the pseudonymous writings of people like Benjamin Franklin. And journalists have largely continued with the founding fathers' charge of maintaining freedom and democracy by helping voters to make informed choices, following the trail of how tax dollars are spent, and covering the dealings of local, state, and federal government. A free press, quite simply, is the foundation of democracy.

Technology, on the other hand, is a tool. The arrival of each new medium over the past century—film, radio, TV, cable TV, internet, World Wide Web, smartphones, apps— has changed the way information is conveyed and consumed. As each new medium appeared, it joined the others, with nearly all collapsing in upon themselves to create the digital landscape that permeates every waking moment of the day. None is recognizable as the distinct technologies they started as.

Trusted, quality content helps us make sense of it all, steering us repeatedly back to the stories that actually affect our lives. But high-quality journalism comes at a high price and, as we've noted previously, today's digital business model favors those who own the technology, notably the Frightful Five.

Chris Anderson pointed to the parasitic, paradoxical relationship between companies that disseminate news across digital media and the journalists who produce it. He speculated that web search behemoth Google has a great deal to lose with the decline of journalism. Online editions have gained eyeballs that consume free news at the expense of print editions, which actually pay for newsgathering. The growth of this not-making-any-money model translates into an ever-changing print and broadcast industry that can afford correspondingly fewer reporters. Anderson noted:

> That means there is less local news for Google to index. There may be more local information, but it can no longer use the fact that it came from a professional news organization as an indicator of quality. Instead, it has to figure out what's reliable and what's not itself, which is a harder problem. So Google would very much like the newspapers to stay in business, even as the success of its own advertising model in taking that market share away from them is making that more difficult.[29]

And digital platforms aren't the only ones that need journalism to survive. Without journalists to drag our attention back to what's actually important to our country and communities, democracy may not have a chance.[30]

Trust as a Hallmark of Journalism

The wealth of information resources on the web and constant engagement of social media by the public means only a small window of time remains when journalists can intervene and help guide the conversation.

Irish journalist and author Mark Little noted that it there is a "golden hour" between "the time it takes social media to create either an empowering truth or an unstoppable lie, when a celebrity death trends on Twitter, or an explosive video surfaces on YouTube. In other words, when journalism can matter most."[31]

The role of journalists increasingly is that of an information tour guide, not just a filter, but a *trusted* filter, to break through the noise of witnesses and amplifiers who share information across social networks. **Amplifiers** are people with wide social networks—many friends and followers—who have

---

29   Anderson (2009), pp. 122–123.
30   Portions of this chapter were originally published in chapter 3, "The evolution of media technologies," of *Media Smackdown: Deconstructing the News and the Future of Journalism* by Abe Amidor, Jim Kuypers, and Susan Wiesinger. 2013. New York: Peter Lang Publishing, Inc.
31   "Finding the wisdom in the crowd," by Mark Little, *Nieman Reports*.

gained a degree of trust and respect. When they post something, people may give that information more weight, causing it to spread quickly. But that doesn't mean amplifiers are purveyors of accurate information, just that their content is more likely to be seen, believed without fact checking, and shared.

Journalists must rely not only on the immediacy afforded by social media sites like Twitter, but on verification and accuracy—good information amid the retweets and gossip—in order to ensure that journalism retains its value in a digital world.

A primary reason journalism has struggled in its transition to digital is its underlying mission: Protect the democracy by serving the public.

A journalist has a stake in the outcome of every story that affects a community, as community building is an integral part of a journalist's job. Low voter turnouts? Few people running for office? Low-to-no attendance at important meetings of local government? Declining readership? The Daily Me? Fail for journalism, fail for the community.

Remember the weapon analogy used in Chapter 2 to discuss the five waves of digital media? The closest that journalists should get to personalization is at the level of a targeted audience (i.e., the shotgun). And a targeted audience is always focused on community, not individuals.

Narrowing digital content down to a demographic level weakens community, and hyper-personalization has the potential to destroy it. In short, it's a journalist's job to say, "Look over here! Something of real importance to your life is happening, and here's what you may be able to do about it."

When journalists forget their mission and attempt to lure in individuals by tempting them with shiny bits of entertainment and drama, journalism fails. And when journalism fails, democracy fails.

Summary

As social media platforms have grown in popularity and evolved technologically, their use has consequences for civil discourse and the health of democracies. The relatively new phenomenon of U.S. presidents using social media to speak directly to platform users has consequences including exclusion of non-users, avoidance of context and interpretation by journalists, and increased ideological division caused by communication primarily those who share the administration's perspectives. Social media companies have enacted policies that actively protect the speech of some users over others, largely those with significant followings. With the deplatforming of then-President Donald Trump from major social media platforms, it has become evident that the power to control speech is held by a few increasingly powerful companies, whose advertising practices have significantly weakened the country's free press.

Key concepts

Amplifiers
Deplatforming
Facebook's XCheck
Jeffersonian Scenario
Journalism
Journalists
Media
"The media"
Medium
Pandora Scenario
Pangloss Scenario
Technology
Tenets of democracy
Three scenarios for the future of technology and a strong democracy

References

Allyn, B. (2021, April 5). Justice Clarence Thomas takes aim at tech and its power "to cut off speech." *NPR*. Retrieved from https://www.npr.org/2021/04/05/984440891/justice-clarence-thomas-takes-aims-at-tech-and-its-power-to-cut-off-speech

Anderson, C. (2009). *Free: The past and a future of a radical price*. New York, NY: Hyperion.

Barber, B. R. (1998, May). Which technology and which democracy? *Democracy and Digital Media Conference, MIT*. Retrieved from http://web.mit.edu/m-i-t/articles/barber.html

Barber, B. R. (2006). Pangloss, Pandora, or Jefferson? Three scenarios for the future of technology and strong democracy. In R. Hassan & J. Thomas (Eds.), *The new media theory reader* (pp. 188–202). New York, NY: Open University Press.

Bermingham, P. (2021, January 11). Merkel among EU leaders questioning Twitter's Trump ban. *Politico*. Retrieved from https://www.politico.eu/article/angela-merkel-european-leaders-question-twitter-donald-trump-ban/

Breton, T. (2021, January 10). Capitol Hill—the 9/11 moment of social media. *Politico*. Retrieved from https://www.politico.eu/article/thierry-breton-social-media-capitol-hill-riot/

Cherlin, R. (2014, August 4). The presidency and the press. *Rolling Stone*. Retrieved from https://www.rollingstone.com/politics/politics-news/the-presidency-and-the-press-74832/

Confessore, N. (2018, April 4). Cambridge Analytica and Facebook: The scandal and the fallout so far. *The New York Times*. Retrieved from https://www.nytimes.com/2018/04/04/us/politics/cambridge-analytica-scandal-fallout.html

Dwyer, D. (2011, February 14). Obama's media machine: State run media 2.0? *ABC News*. Retrieved from https://abcnews.go.com/Politics/president-obama-white-house-media-operation-state-run/story?id=12913319

Eilperin, J. (2015, May 26). Here's how the first president of the social media age has chosen to connect with Americans. *The Washington Post*. Retrieved from https://www.washingtonpost.com/news/politics/wp/2015/05/26/heres-how-the-first-president-of-the-social-media-age-has-chosen-to-connect-with-americans/

Hatmaker, T. (2021, March 25). Zuckerberg blames Trump, not Facebook, for the Capitol attack. *TechCrunch*. Retrieved from https://techcrunch.com/2021/03/25/facebook-capitol-riot-zuckerberg-sandberg-comments-hearing/

Lerman, R. (2021, January 20). Biden takes over POTUS Twitter account, inheriting a blank slate from Trump. *The Washington Post*. Retrieved from https://www.washingtonpost.com/technology/2021/01/20/potus-twitter-account-biden-trump/

Little, M. (2012). Finding the wisdom in the crowd. *Nieman Reports, 66*(2), pp. 14–17. Available online at https://niemanreports.org/articles/finding-the-wisdom-in-the-crowd/

Marineau, S. (2020, September 29). Fact check US: What is the impact of Russian interference in the US presidential election? *The Conversation*. Retrieved from https://theconversation.com/fact-check-us-what-is-the-impact-of-russian-interference-in-the-us-presidential-election-146711

Perrett, C. (2020, June 6). Trump broke his all-time tweeting record amid nationwide protests. *Insider*. Retrieved from https://www.insider.com/trump-breaks-record-most-tweets-in-a-single-day-2020-6

Rosenberg, M. & Dance, G. J. X. (2018, April 8). "You are the product": Targeted by Cambridge Analytica on Facebook. *The New York Times*. Retrieved from https://www.nytimes.com/2018/04/08/us/facebook-users-data-harvested-cambridge-analytica.html

Treisman, R. (2019, October 25). As President Trump tweets and deletes, the historical record takes shape. *NPR*. Retrieved from https://www.npr.org/2019/10/25/772325133/as-president-trump-tweets-and-deletes-the-historical-record-takes-shape

Twitter deletes new Trump tweets on @POTUS, suspends campaign account. (2021, January 8). *Reuters*. Retrieved from https://www.reuters.com/article/us-usa-election-trump-twitter-removal/twitter-deletes-new-trump-tweets-on-potus-suspends-campaign-account-idUSKBN29E02H

CHAPTER 9

It Really Is a *Thing:* The Internet as Infrastructure

Figure 9.1. The internet—perhaps the most kick-ass communication invention ever—was invented by DARPA to enable real-time sharing of files and programs. Planning of the internet started in 1962, four years after the U.S. government created DARPA, and the network, then called ARPANET, was officially up in 1969. Doonesbury © 2013 G. B. Trudeau. Reprinted with permission of Andrews McMeel Syndication. All rights reserved.

Takeaway: The internet and World Wide Web are not the same thing. While the internet comprises a physical infrastructure, the web is just one of many subnetworks that function within it.

The internet is a computer network that allows real-time file sharing. And if the internet is a computer network, then the World Wide Web clearly must be a *worldwide* computer network. (Sounds right enough, even if it's *wrong*.)

While the terms often are used interchangeably, the internet and web are not the same things.

There's a popular origin story for the internet that attributes its invention to the U.S. military and implies that it was created as a weapon. The former is true, while the latter probably is closer to how the internet is used by countries worldwide today as a tool for cyberwarfare.

The internet's roots were set by Defense Advanced Research Projects Agency (DARPA),[1] a well-funded branch of the U.S. Department of Defense, that created a prototype for computers networked via the telephone lines called ARPANET.[2] The first successful trial of the network was in 1969, connecting research teams at University of California, Los Angeles and Stanford Research Institute.

Other researchers worldwide were coming up with their own computer network protocols, but a key problem emerged: They were closed systems that couldn't communicate with each other. The solution was the invention of Transmission Control Protocol/Internet Protocol, or TCP/IP, which allowed connection of networks.

TCP/IP is the set of standardized protocols that comprise today's internet, allowing an expansive network of computers to share information in real time. Its blueprint was created in 1973 by a team of DARPA scientists led by Vinton Cerf and Robert Kahn,[3] who share the title of fathers of the internet.

The internet became part of American daily life after the arrival of personal computing in the mid-1980s. This is when AOL, CompuServe, and Prodigy, among others, established their early dominance as online service providers. What was the service? Email, chat rooms, and bulletin boards—places to talk to people online, using a text-based interface that was accessed via dial-up modems.[4] Not only did dial-up tie up your phone line, but it also was very, very slow.

British computer scientist Tim Berners-Lee, who created HyperText Markup Language, invented the web in 1989. **HTML** does three things:

1. Provides a universal language that can be read by all computers, eliminating a need for platform-specific software for different computer operating systems (e.g., Microsoft Windows versus Mac OS);
2. Tells a browser how a web page should look; and,
3. Facilitates hyperlinks and hypertext, which link web pages directly to other web content using encoded web addresses. Here's a brief explanation of how this works:
   - A **Uniform Resource Locator,** which is a web address, may be hyperlinked. For example, http://www.theguardian.com is the website for *The Guardian,* a Pulitzer-prize-winning, internationally focused news organization.

---

1   Worth noting here that DARPA researchers were responsible for creating the first computer mouse, a personal computing prototype called Project MAC, video teleconferencing, GPS, and more—all in the 1960s. Check out the "About Us" tab at darpa.mil to learn more about the technological innovations that came from U.S. government work.

2   Short for Advanced Research Projects Agency Network.

3   You likely know the names of Jeff Bezos, Bill Gates, Steve Jobs, Elon Musk, and Mark Zuckerberg, and can identify the companies they created. Strange that you may never have learned the names of the two researchers who created the network that virtually all digital communication runs across …

4   If you've never heard dial-up, put "dial-up modem sound" into a search engine and listen. *That* was the sound most people heard for nearly two decades that meant they were connecting to the internet.

- **Hyperlinking** allows a user to click on an encoded URL and leap to another page. A "live" hyperlink generally is underlined and of a different color from surrounding text. A hyperlink can also be a full URL, which generally automatically links in things like email when you put a space beyond it, like https://www.theguardian.com/.
- **Hypertext** is linked words on a page that explain where the link is sending you, which a URL does not necessarily do. If the words "*The Guardian*" were highlighted with color and underlined on a web page, you would know that you could click on them to go to a different site or a particular page within the site.

The web itself is a subnetwork of the internet, underpinned by hypertext transfer protocol, or **HTTP**—the standardized protocol that comprises the World Wide Web.

You can imagine it like this: If the United States is the web, then the internet is the world; the web is just one subnetwork of many that rely on the internet. Others include email, instant messaging, multiplayer gaming, video streaming, and digital radio. For example, email functions via IMAP, POP, or SMTP—Internet Message Access Protocol, Post Office Protocol, Simple Mail Transfer Protocol—all of which are protocols that run on the internet.

So the internet and web are distinct inventions, TCP/IP and HTTP, one built upon the technology of the other. How do those languages translate into web pages? The simple answer is that all web pages must be accessed via **browsers** (such as Opera, Firefox, Chrome, and Safari), which decode digital signals into words, images, motion, and interactivity.[5]

Web content is discovered by web crawlers (aka spiders) that read HTML code, with key information stored and indexed by **search engines,** like Google, Yahoo!, Bing, OneSearch, Ask, and DuckDuckGo.[6]

Thinking beyond the screen(s) in front of you, how does all this information move from place to place? What does the internet look like? How would you describe its structure? Where *is* it? If you're waving your arms around you in an "it's out here" motion, you're sort of right and completely wrong.

Figure 9.2. What does the internet look like? At left, a mapping of the internet created by The Opte Project,[7] with the most active interchanges showing up brighter and looking somewhat like dandelion seed parachutes; center, a dandelion blowball; at right, the Southeastern United States at night from space.[8] The structure of the internet doesn't have the uniformity of the blowball or the unevenness of lights from space. It kind of resembles a hybrid of the two.

5   It's an incredibly complex process that is far beyond the scope of this book and deserves the focused attention it is given by a wide range of platform-, application-, and interface-specific classes and books.
6   It's possible you weren't aware there were other search engines than Google, but there are actually quite a few. DuckDuckGo, for example, sells itself as the pro-privacy solution to Google's continual surveillance. Twitter and TikTok have grown in popularity as quasi-search engines—a Wikipedia-ish, user-generated approach to search.
7   http://www.opte.org/the-internet/
8   http://earthobservatory.nasa.gov/IOTD/view.php?id=79800

The Internet as Infrastructure

Wireless devices, like your smartphone, TV, tablet, laptop, and any other Wi-Fi-connected device, are receiving signals from wired sources. There's an interesting history to this.

Telephones started out as hardwired devices—phones were hooked to phone lines and had cords connecting the handset to the telephone. Radio and television started out as wireless, receiving their signals through the air.

This is strange, if you think about it. It makes sense to have a telephone you can take with you. And limiting radio and television to what could be received wirelessly—via analog bandwidth—resulted in government regulation and programming domination by a few networks.

MIT Professor Nicholas Negroponte suggested in the 1980s that wired phones and wireless broadcasting needed to trade places in terms of transmission. That idea today is known as the **Negroponte Switch**.

We've reached a point in time where that switch has occurred—when most television sets are hardwired and the majority of phones operate wirelessly via signals sent to and from cellphone towers planted strategically throughout our current technology landscape.

But what about TVs and smartphones that rely on Wi-Fi to stream content and communicate via the internet? While both are using signals, not direct hardwiring to send and receive content, the presence of Wi-Fi means that a physical internet connection exists nearby.

That connection is directly to a **Wi-Fi** router or transmitter, which allows wireless data transfer between broadband or fiber optic internet and Wi-Fi-enabled devices.

Worth noting here that Wi-Fi, which was invented in 1997, doesn't actually stand for anything. Turns out the Wireless Ethernet Compatibility Alliance, the group that created the common set of standards that make wireless local area networks possible, hired a marketing company to come up with a brand that sounded sexier than IEEE802.11 Direct Sequence.[9]

Whether hardwired or connected via Wi-Fi, digital devices are tapping into the physical system of cables, towers, and satellites that is the internet. In many ways it's a technological mash-up, with newer stuff and older stuff carrying content all over the globe. The cables running under the ocean from continent to continent, however, are as tangible as the bridge that allows you to drive over a river.[10]

Underwater communication cabling was first placed in the mid-1800s to facilitate the transatlantic telegraph system. The cable system has been funded since its inception by private companies, which likely resulted in more expedient cable laying than would be possible if international accords had to be brokered among governments.

Today there are more than 400 cables in use, spanning more than 800,000 miles.[11] That's more than three times the distance from Earth to the Moon.

The cables that cross the world's oceans today carry bundles of fiber optics, which allow near-instant transmission of information from one continent to another. Earth-orbiting satellites are another means of international communication, but they often have delays in information conveyance, known as **latency.**[12] Despite the availability of satellites, undersea cables still transmit 99% of international communication traffic.[13]

---

9   Info from alliance founding member Phil Belanger, who was part of the name selection process. In "WiFi isn't short for 'Wireless Fidelity'" by Cory Doctorow, BoingBoing.

10 For the fully interactive version and more information about the submarine cable system, go to https://www.submarinecablemap.com.

11 For current numbers, put the terms "submarine cable 101" and "telegeography" into your favorite search engine.

12 Satellite latency is measured in hundreds of milliseconds. It's most noticeable when watching a televised interview being conducted via satellite: The interviewer asks a question and there is a brief time delay between the question being asked and the interviewee receiving it. It's one of those rare tech examples that draws attention to the ability of information communication to cross time and space.

13 Telegeography.com is one of the best sources online for information about the physical backbone of the internet.

And, while companies like Alphabet, Meta, and Microsoft have significant investment in undersea cable ventures, they're also part of a very expensive race to invent something that reduces that need and expands digital communication connectivity to parts of the world that lack it.

In the meantime, global internet access, speeds, and innovation vary widely.

Let's think about the U.S. for a moment: Where do you think it ranks globally with internet download speeds?

When it comes to fixed broadband—hard-wired access to homes, schools, government, and businesses—the U.S. ranks 14th in the world for average download speed. Number one? Chile. For mobile, the U.S. ranks 18th, with United Arab Emirates in first.[14]

Why do speeds vary? Because countries make different decisions regarding the management of internet infrastructure (which is covered in more depth in Chapter 13). For example, the U.S. has relegated internet infrastructure development and maintenance largely to for-profit corporations.

Harvard law professor Susan Crawford has long argued that the U.S. is falling behind in internet infrastructure.

> That's the odd thing about internet access in America: it's always in the news, with daily stories about the wonders of augmented reality, the Internet of Things, and driverless cars. No one seems to stop to ask whether those advanced uses of data will work reliably, where and for whom.[15]

The problem is largely a result of internet infrastructure decision-making being relegated to states and municipalities, with most internet access provided by corporations, like Xfinity and AT&T. As a result, internet is fastest in wealthier areas in cities and suburbs, and slowest in rural areas, in inner cities, and on tribal lands, where these companies have little incentive to extend service. That's because there's not as much money to be made.

This means that fiber-optic internet, which provides the fastest speeds, has not been evenly distributed across the country. Instead, many people still have broadband and pay far higher rates for far slower service than in other places in the world. As Crawford noted,

> Today in America, local cable or telco monopolies, uncontrolled by either competition or oversight, can charge whatever they want for whatever level of service their shareholders will accept. … This reality is causing problems for our future as an innovative and just country, as we fall further behind in the global race to create new jobs and new ways for citizens to make choices with their lives. We are amplifying and entrenching existing rural/urban divides and, even more starkly, inequality of opportunity.[16]

The result is kind of a mixed reality, where those engaged in the digital conversation perceive one world, while others who have slower or non-existent internet service perceive another.

Disruption to the Internet's Infrastructure

The internet began—and remains—a cobbled together system of individual computers and computer networks. By its very nature it's decentralized, which means that there is no central control or monitoring.

There are hundreds of millions of internet connections that send digital signals to and receive them from routers, data centers, exchange points, nodes, supernodes, and hubs across the planet. That's part

---

14   For current rankings, search online for the Speedtest Global Index. While you're there, take advantage of Speedtest tools to measure the internet speed on your own device.
15   Crawford (2018), p. 37.
16   Ibid., p. 47.

of what makes the internet's infrastructure so robust: If one section fails, there are numerous ways information can be routed around the damage.

Unlike an invention like the lightbulb, where you can rather quickly assess its function and purpose, the internet is everywhere, while at the same time invisible and elsewhere. It's extremely complex and incredibly simple, all at the same time.[17]

Somewhat like a city road system, there are the equivalents of residential streets, feeders, collectors, and arterials that carry varying loads of traffic, as well as freeways that bypass large sections for faster travel. If one route is reduced or eliminated, traffic is detoured to another route toward the same destination.

Now—keeping in mind that this road system is our mental model of the internet—imagine that a major accident has reduced traffic lanes on the freeway and vehicles must be detoured onto surface streets—and it's rush hour. Some people are stuck on the freeway and are barely moving at all, while everything slows on every road that has to absorb the extra traffic.

Ugh.

So, parts of the internet can and do fail for an extensive range of reasons.

What Happened in Nashville Didn't Stay in Nashville

Early on Christmas Day 2020, downtown Nashville, Tennessee, was rocked by a massive explosion originating in an explosives-filled motorhome.

The target?

A multi-story AT&T service center that provided telecommunications to a regional swath of individuals, homes, and businesses in Tennessee, Kentucky, Alabama, and Georgia. The service center sustained significant damage, as did more than 60 other downtown buildings. Due to the timing, and a warning blaring from the motorhome 15 minutes prior to the blast, most were unoccupied or evacuated.

The motive?

Absolute certainty of "why" died with the single suspect, but investigation revealed he was plagued by paranoia and immersed in online conspiracy theories that spread misinformation about 5G cellular technology, among other things.

The blast knocked out internet, landline, cell phone, 911, air traffic control, and fire alarm services, as well as access to credit card terminals. Most were restored within a few days, but the attack raised concerns about the vulnerability of U.S. telecommunication infrastructure.

The catastrophic failure of a single communications hub is what computer science professor Douglas Schmidt called the "Achilles' heel" of the system:

Having a critical facility in a major metropolitan area next to a street without any other protections than a thick wall is crazy. ... This is a wake-up call that, if people treat it right, will help with future situations and be better prepared.[18]

As the world moved online amid pandemic lockdowns in 2020, there was significant concern that the internet infrastructure would not be able to handle the sudden increase in bandwidth-intensive

---

17   "How the internet was invented" by Ben Tarnoff, The Guardian.
18   "Nashville bombing froze wireless communications, exposed 'Achilles hell' in regional network" by Yihyun Jeong & Natalie Allison, *Nashville Tennessean*.

video teleconferencing needed for … nearly everything. There was also the reality that people had more time to watch streaming video, listen to streaming audio, and game—activities that consume a lot of data via the internet. Rather than streaming four hours in the evening, people were suddenly streaming across multiple devices for up to 10 hours per day.[19]

Governments, particularly those in Europe, were concerned the internet would collapse, adding to the looming pandemic chaos. In response, streaming video providers "throttled" such leisure-based content to reduce the load on the internet infrastructure. This included Netflix, Amazon Prime Video, Disney+, and YouTube, which defaulted to standard-definition, rather than high-definition video, in an effort to reduce their internet traffic by 25%.[20]

Internet Service Providers reported increased traffic worldwide, but those managing big chunks of internet connectivity said the internet's decentralized structure would help it absorb the sudden surge. Cisco's chief technologist, Chintan Patel, described the system like this:

> The internet is not just one thing. It's like a living breathing human with lots of different neurons and connection points working to keep it alive. Even if you cut off one finger, the rest of the body will keep it alive.[21]

A couple of years later, Patel's analogy proved apt when the volcanic eruption on Tonga severed the *one* undersea cable connecting the island nation to the world. A piece failed, the rest carried on.

While there are and will continue to be frequent examples of internet and technology failure, following are a few classics—disruptions that emphasized the fragility of the physical system.

In December 2006, an earthquake struck off the coast of Taiwan, damaging and/or severing nine submarine cables. This didn't cause a sudden halt to all international communications between Asia and the United States, but it did disrupt internet and telephone service to millions of people in dozens of countries. Some locations lost service completely, while others found international communication slowed as traffic was rerouted. And it took more than a month to repair the lines.

Just over one year later, three submarine cables were cut over the course of three days in the Middle East. The first two were severed in the Mediterranean Sea between Egypt and Italy, with the other cut near Dubai.

This human-caused damage—all three were attributed to ship anchors—stopped or slowed internet access for millions of people in Europe, Egypt, Saudi Arabia, Pakistan, India, North Africa, Singapore, and Jakarta. In fact, the first two cables cut stopped three-quarters of the international internet traffic between Europe and the Middle East until they were repaired two weeks later.

There are other cases of internet disruption that amount to sabotage, sometimes by governments and sometimes by political activists. These disruptions are less about damage to the physical system and more about messing with programs that route traffic through it and the electricity supply that powers it.

In early 2011, a veritable flood of social media postings on Facebook, YouTube, and Twitter drew international attention to civil unrest in Egypt. The government launched a counterattack, doing something that no one thought possible: It flipped the "off" switch for the internet.[22]

---

19   "The internet is under huge strain because of the coronavirus" by Ryan Browne, CNBC.
20   "Disney+ joins Netflix, Apple, Amazon and YouTube in throttling streaming quality during coronavirus outbreak" by John Archer, Forbes.
21   Browne (2020), op. cit.
22   Speculation is that someone with a strong understanding of foundational internet architecture guided the government in powering down key servers or changing routing programs. Either way: no internet.

For five days, a blocked Internet **choke point** isolated Egypt's more than 25 million internet users from the rest of the world. This didn't stop the revolution, but it did briefly prevent it from being posted on social media.

Choke points are centralized pathways on the Internet where large amounts of information are channeled.

Think back to the road system analogy several paragraphs back. Now imagine three major north-south routes across your community are completely blocked, as well as three major east-west thorough-fares. How are you going to get across town without crossing any of those streets? If you know your community well, you may be able to navigate the surface streets. But you'll be hampered by a whole bunch of confused drivers who are blocking traffic. It's kind of like citywide gridlock.

A shutdown similar to the one in Egypt happened in spring 2014 when the Turkish government blocked Twitter access by targeting a particular choke point, the country's **Domain Name System** server.

Shutting down the DNS effectively blocked the internet to more than 36 million Turkish users, or nearly 46% of the country's population.[23] In response, Turk protesters launched a graffiti campaign that directed people to reconfigure computer Internet Protocol setup to DNS 8.8.8.8 and 8.8.4.4, which would reroute internet traffic to Google's DNS servers and restore service.

In both the Egyptian and Turkish choke point shutdowns, the internet was *confused*. Traffic didn't know where to go with central pathways blocked.

Why did these efforts work? It was largely because the Egyptian and Turkish governments own the systems that connect these countries to the internet and, as a result, have unfettered access to it—as is common with many authoritarian governments.

The Weakest Link

The Domain Name System (DNS) is widely recognized as being the least secure part of the internet and therefore among the most likely to be exploited by hackers.

Every electronic device with an operating system—from computers to tablets to smartphones—has a unique IP address that conveys a variety of information, from who provides your Internet service to a rough approximation of the physical location of the device.

To discover your IP address and see what information is available about it, put the term "IP address lookup" in a search engine.

Rerouting your device's DNS is relatively simple: Go to system preferences, network, and look for TCP/IP, IPv4 address configuration. That's where you can manually input the DNS server information. You also can go to DNS and add an IPv4 address (or two).

But you really shouldn't do this without a reason. Please learn more before you start messing with your IP address. Just a suggestion …

Could the same thing happen in a much larger country, say the United States? Perhaps not a national outage, but it is possible that large areas of the country could suffer a tech blackout or data interruption. Something as simple as a misconfiguration of coding in a routing table could affect access to particular platforms.

---

23   For an excellent source of internet usage statistics worldwide, go to http://www.internetworldstats.com/.

Routing tables are what guide information traffic through the internet system. Think of it like the barcodes shipping companies use to track packages: The package doesn't seek out the best route to where it's going; the barcode guides the package through the system. Corrupt data (or smeared ink) in a barcode could have your package bound for Lafayette, LA, rather than Lafayette, CA. And it wouldn't be able to find its way back without human intervention.

That's essentially what a damaged routing table could do to a whole lot of information flowing through the system—it would be sent on its way but not arrive where it was intended to go. And routing outages can and do happen.

In 2021, all of Meta went down globally for five hours after a DNS configuration change (see DNS sidebar for more on what this technology does). The outage affected *billions* of Facebook, Instagram, What's App, and Oculus users, as well as Meta's internal communications and the employee badges needed to access offices and servers to fix the problem. Keeping in mind that people use Meta platforms to sign into other websites, the outage also prevented that access. In 2019, another routing table change took the same Meta platforms offline for 24 hours.[24]

As the Meta examples illustrate, systems that share routing configurations across platforms are prone to widespread outages when something goes wrong.

And we're not picking on Meta.

In 2020, one U.S. region of Amazon Web Services, one of the largest cloud computing hosts in the world, went down for several hours, taking with it dozens of apps, services, and websites, including some Adobe products, Roku, *The Philadelphia Inquirer*, Pocket, *The Washington Post,* and WNYC.

Just a few weeks after the AWS outage, a huge swath of Alphabet platforms went down at once globally, for about an hour. This included Gmail, Google Docs, YouTube, Google Maps, Google Pay, Google Home, and Chromecast. Google-as-search-engine and Chrome remained online, albeit moving very, very slowly. TechCrunch called the outage "an alarming reminder of just how far Google reaches, and how many of our services—productivity, entertainment and home/utility—are tied up with a single, proprietary provider."[25]

Beyond platforms, control of internet access is a significant issue. Consolidation of technology ownership therefore is a relevant concern (see Chapter 3), because we're approaching the point where Xfinity (Comcast) and AT&T are the only Internet Service Providers for the majority of the U.S. population. And both also offer digital phone services, meaning people could also lose not only internet, but phone service in the event of a natural or human-generated crisis.

Not only could a major ISP outage disrupt the flow of internet into homes, schools, government, and businesses nationwide, but those same ISPs control access to the conduit that distributes digital content. Imagine if one of those few providers opted to block access to—cancel—an entire social media network. After all, this is what Alphabet and Apple did when they deplatformed the social media network Parler in January of 2021. Now imagine Xfinity, Verizon, or AT&T deplatforming Facebook, YouTube, TikTok … Stranger things have happened.

With all that being said, a natural disaster—such as a significant earthquake in the Columbia Gorge that interrupts power to even one of the gargantuan data servers in the United States—could quickly make things interesting for the internet-reliant world.[26] Just a portion of one massive data center going offline can interrupt parts of the World Wide Web for millions of people worldwide—and has done so with increasing frequency.

---

24   "Facebook is back online after a massive outage that also took down Instagram, What's App, Messenger, and Oculus" by Richard Lawler and Alex Heath, *The Verge.*
25   "Gmail, YouTube, Google Docs and other services go down in multiple countries" by Ingrid Lunden, *TechCrunch.*
26   For a comic view on what a widespread internet outage in the United States might be like, watch "Over Logging," Season 12, Episode 6, of Comedy Central's *South Park.* First aired April 16, 2008.

Like other infrastructure systems—from roads, to electricity, to water supply—the physical structures of the internet can, and do, fail. And they fail for the same kinds of reasons: age, accidents, natural
disasters, and sabotage.

In fact, internet outages occur every minute of every day, everywhere.[27] Just put "internet outage"
into a search engine to see the range of causes, from thunderstorms to cows eating the cable lines. We
kid you not.

What the Decentralized Structure of the Internet Hath Wrought

The internet began—and remains—a cobbled together system of individual computers and computer
networks. By its very nature it's decentralized, which means that there is no central control or monitoring. It's been termed a "future-proof" technology, in that it's been a stable base to share content across
and build applications upon for nearly 50 years.

Harvard professor Jonathan Zittrain pointed to the internet as "a collective hallucination that
functions because millions of people and companies believe in it."[28] And it likely couldn't and wouldn't
be built the same way today. As Zittrain noted,

> Rather than a single centralized network modeled after the legacy telephone system, operated by a gov-
> ernment or a few massive utilities, the internet was designed to allow any device anywhere to interop-
> erate with any other device, allowing any provider able to bring whatever networking capacity it had to
> the growing party. And because the network's creators did not mean to monetize, much less monopo-
> lize, any of it, the key was for desirable content to be provided naturally by the network's users … So
> the internet was a recipe for mortar, with an invitation for anyone, and everyone, to bring their own
> bricks.[29]

The idea that the internet is beyond control of any one government, organization, or company
lends it a remarkable resilience. But the Frightful Five have taken control of huge swaths of the internet, creating profit-centered, closed networks that allow companies to decide what—and who—stays
and goes.

Contemporary internet, web, and platform content contains not only a wealth of information and
means of connection, but also individual harassment, inadvertent misinformation, and intentional
disinformation designed to sow division. Beyond that, the ability to share, edit, alter, move, and delete
content—collectively causing content drift and link rot on the internet and World Wide Web—means
that the technology's very design is contributing to unprecedented gaps in the historical record.

Zittrain and two other Harvard researchers conducted a study to see how things like hyperlinks
embedded in Supreme Court opinions since 1996 had stood the test of time. They found that 50% of
the links no longer worked.[30]

In one rather amusing example of **link rot**—essentially broken hyperlinks on a webpage—an
opinion by Justice Samuel Alito included a link to a site that no longer existed. Instead, the domain
had been purchased by someone else and in place of the cited link's content was a meme-like message
"about the transience of linked information in the internet age."[31] With Alito's opinion still linked to
the same page, the message from the new management has evolved from a statement about link rot into
what's termed the political rot of the Supreme Court.[32]

---

27   Downdetector.com is a great place to start if you think a platform is down (and still have access to the internet …).
28   Zittrain (2019), p. 369.
29   "The internet is rotting" by Jonathan Zittrain, The Atlantic.
30   Zittrain, Albert, and Lessig (2014).
31   An academic program called Perma.cc makes a backup and permanent link to avoid this very problem. Here's the Perma.
     cc record for the Alito opinion error message: https://perma.cc/0gwuqRxEJJW?type=image.
32   Hard telling what's there now, but you can check it out at ssnat.com.

Zittrain participated in a second study that looked at two million articles from *The New York Times*. They found that more than half the articles since 1996, when the NYT first went online, had at least one broken link. Other studies found that 75% of the academic science, technology, and medicine articles put online since 1994 had at least one broken link.[33]

If you've ever heard the phrases, "there's a tweet for that" or "what goes online, stays online," Zittrain pointed to the counter-phenomena of personal online content, rumors, and alternative facts effectively *never* going away.

> People communicate in ways that feel ephemeral and let their guard down commensurately, only to find that a Facebook comment can stick around forever. The upshot is the worst of both worlds: Some information sticks around when it shouldn't, while other information vanishes when it should remain.[34]

In the past, documentation of human history was relegated to stored paper, ideally stored in multiple locations like libraries. It could be difficult to access an original document, much less alter it without leaving evidence you had done so. Zittrain's chief concern is that the transition to a relatively new and ever-evolving technological landscape has the capacity to wipe out chunks of historical documentation very quickly.

It's hard to deny the similarity to the world imagined in George Orwell's *1984*, in which "facts" were in continual flux, based on need. Instead of the government controlling the ever-changing narrative, it's the commercial applications that dominate digital technology that have infiltrated every part of our daily lives. In this twist on Orwell's vision, the network's very resilience takes on a cockroach-like quality: the internet could very well contribute to and survive the collapse of civil society, much as a disease-carrying cockroach might contribute to and survive the collapse of humanity.

And, what if the internet isn't as resilient as it appears to be? Could the physical infrastructure collapse? Given contemporary reliance on undersea cables for 99% of data transmission across the globe, scientists say yes. Those cables are outfitted with repeaters to ensure that data makes its way intact, and those repeaters could be knocked out by geomagnetic currents.[35]

The culprit likely would be a solar storm—technically known as a coronal mass ejection—which could also affect electric grids and satellites. While broadband and fiber optic cables would not be affected and the power likely would return quickly, the global connections that the internet facilitates might not. Tech journalist Lily Hay Newman, after interview with a computer science researcher, framed it like this:

> A solar storm that disrupted a number of these cables around the world could cause a massive loss of connectivity by cutting countries off at the source, even while leaving local infrastructure intact. It would be like cutting flow to an apartment building because of a water main break.[36]

While it's not likely that all submarine cables would be affected by a solar storm, the problem takes us back to the idea of choke points. There are parts of the internet that carry more traffic than others and a collapse of key routes would cause the DNS system to go awry. As Hay Newman concluded, "it's the internet version of traffic jams that would happen if road signs disappeared and traffic lights went out at busy intersections across a major city."[37]

---

33   Zittrain (2021), op. cit.
34   Ibid.
35   "A bad solar storm could cause an 'internet apocalypse'" by Lily Hay Newman, *Wired*.
36   Ibid.
37   Ibid.

So, fun note to end this chapter on, eh? That's your primer on the internet. There's far more to know, obviously; our goal has been to highlight the basics.

And it's probably as good a time as any to offer an apology to those of you who may have just discovered that discussion of the systems and ownership of the technology we use can trigger a bit of technophobia.

But you've made it this far and may be relieved to know that we're now going to talk about the World Wide Web, which was invented for a much different purpose than the information transport model of the internet.

Summary

The term "online" may be used to describe anything from email to websites to smartphone apps, but the relationship among them is more complex. All involve distinct protocols. The internet—TCP/IP—is a vast network underpinned by an infrastructure of computers, data centers, cables, towers, and satellite systems. The World Wide Web—HTTP—is a subnetwork of the internet, invented by Tim Berners-Lee in 1989. The invention was HyperText Markup Language, which is better known as HTML. The infrastructure of the internet can be physically damaged, as well as hacked, using a variety of means that confuse the flow of data through the system (i.e., disrupt communication via the web, cellphone service, and mobile apps).

Key concepts

Browsers
Choke point
Domain Name System
HTML
HTTP
Hyperlinking
Hypertext
Latency
Link rot
Negroponte Switch
Search engines
TCP/IP
Uniform Resource Locator
Wi-Fi

References

Archer, J. (2020, March 21). Disney+ joins Netflix, Apple, Amazon and YouTube in throttling streaming quality during coronavirus outbreak. *Forbes*. Retrieved from: https://www.cnbc.com/2020/03/27/coronavirus-can-the-internet-handle-unprecedented-surge-in-traffic.html

Browne, R. (2020, March 27). The internet is under huge strain because of the coronavirus. *CNBC*. Retrieved from: https://www.cnbc.com/2020/03/27/coronavirus-can-the-internet-handle-unprecedented-surge-in-traffic.html

Crawford, S. (2018). *Fiber: The coming tech revolution—and why America might miss it*. New Haven, CT: Yale University Press.

Doctorow, C. (2005, November 8). WiFi isn't short for "Wireless Fidelity." *BoingBoing*. Retrieved from https://boingboing.net/2005/11/08/wifi-isnt-short-for.html

Earth observatory. (n.d.). *NASA*. Retrieved from http://earthobservatory.nasa.gov/IOTD/view.php?id=79800

Hay Newman, L. (2021, August 26). A bad solar storm could cause an 'internet apocalypse.' *Wired*. Retrieved from: https://www.wired.com/story/solar-storm-internet-apocalypse-undersea-cables/

Jeong, Y. & Allison, N. (2020, December 29). Nashville bombing froze wireless communications, exposed "Achilles hell" in regional network. *Nashville Tennessean*. Retrieved from https://www.usatoday.com/story/news/nation/2020/12/29/nashville-bombing-area-communications-network-exposed-achilles-heel/4070797001/

Lawler, R. & Heath, A. (2021, October 5). Facebook is back online after a massive outage that also took down Instagram, What's App, Messenger, and Oculus. *The Verge*. Retrieved from https://www.theverge.com/2021/10/4/22708989/instagram-facebook-outage-messenger-whatsapp-error

Lunden, I. (2020, December 14). Gmail, YouTube, Google Docs and other services go down in multiple countries. *TechCrunch*. Retrieved from https://techcrunch.com/2020/12/14/gmail-youtube-google-docs-and-other-services-go-down-simultaneously-in-multiple-countries/

"Over logging" [Television series episode]. (2008, April 16). *South Park* [Season 12, Episode 6]. Comedy Central. Retrieved from http://southpark.cc.com/full-episodes/s12e06-over-logging

The Opte Project. (2021). *The internet 1997–2021*. Retrieved from http://www.opte.org/the-internet/

Tarnoff, B. (2016, June16). How the internet was invented. *The Guardian*. Retrieved from: https://www.theguardian.com/technology/2016/jul/15/how-the-internet-was-invented-1976-arpa-kahn-cerf

Zittrain, J. (2019). 45: Internet. In C. O. den Kamp and D. Hunter (Eds.), *In a history of intellectual property in 50 objects* (pp. 369–375). Cambridge, UK: Cambridge University Press.

Zittrain, J. (2021, June 30). The internet is rotting. *The Atlantic*. Retrieved from: https://www.theatlantic.com/technology/archive/2021/06/the-internet-is-a-collective-hallucination/619320/

Zittrain, J., Albert, K., & Lessig, L. (2014). Perma: Scoping and addressing the problem of link and reference rot in legal citations. *Harvard Law Review Forum*. Retrieved from https://perma.cc/KL84-CEHS

CHAPTER 10

If It's Not the Internet, What Is It? The Web as a Collaborative Tool

Figure 10.1. *From its inception, the World Wide Web had tremendous potential for collaboration, profit ... and distraction. Doonesbury © 1995 G. B. Trudeau. Reprinted with permission of Andrews McMeel Syndication. All rights reserved.*

Takeaway: The original purpose of the web was to serve as a collaborative conduit for intercreative communication.

———————————

There are a handful of well-known names associated with contemporary technological innovation: Sergey Brin and Larry Page (Google), Jeff Bezos (Amazon), Bill Gates (Microsoft), Steve Jobs (Apple), and Mark Zuckerberg (Facebook). Zuckerberg and Jobs each were portrayed as ruthless entrepreneurs in popular movies about their rise from college dropouts to billionaires, so their names may be more familiar to many.

Tim Berners-Lee, inventor of the World Wide Web, has been a far lower-key figure, but it is his work, and the consortium of like minds he built around him, that made the other innovations possible. He began his work on the web in 1989, with it available for public use by 1991. After a period of slow adoption, by the late 1990s the WWW was on the way to achieving its world-wide promise.

Perhaps, if Berners-Lee had stuck with an earlier idea for naming, The Information Mine, we'd more easily remember his contribution to the world. Ultimately, he decided the acronym, TIM, was a tad egocentric and conceptualizing it as a "mine" was not quite in line with his goals. Not only did that not indicate global connection, but "it represented only getting information out—not putting it in."[1]

Berners-Lee documented his efforts to create a society of shared knowledge in the book *Weaving the Web.* It's notable that he begins Chapter 12 with the line, "I have a dream … ."[2]

It takes a bold person to co-opt the phraseology of Martin Luther King, Jr., the best-known civil rights leader in the United States. Of course, it takes an equally bold person to assemble a team and bring the web from dream to reality—forever changing global culture along the way.

To be fair, Berners-Lee's full opening paragraph is, "I have a dream for the Web … and it has two parts." He explains that each is necessary to achieve his vision of the web as "a place where the whim of a human being and the reasoning of a machine coexist in an ideal, powerful mixture."[3]

The two parts of **Tim's Dream for the Web** are:

1. **Intercreativity,** which entails Web users contributing personal expertise and knowledge. They ultimately build something together that is far greater than the sum of its parts.
2. A **Semantic Web,** in which computers do the heavy lifting when it comes to organizing and analyzing content on the Web.
    –   The Web was designed to facilitate machine assistance, via programmed algorithms, which is necessary to manage the sheer volume of content.
    –   Automated Web indexers (aka spiders) continually gather the Web's vast information resources and the search engines synthesize and customize.

How close are we to achieving Berners-Lee's two-part dream? Very.

Creative collaboration today is better known as **crowdsourcing,** where individuals participate by adding their own information, experiences, and expertise. Social media are entirely based on this premise, with users contributing vast information databases and consuming the content.

An early and notable example is Wikipedia, which allows users to contribute information and negotiate reality in the quest for "Neutral Point of View." This means that the content of Wikipedia entries is negotiated and reached via user consensus.

1   Berners-Lee (2000), p. 23.
2   Ibid., p. 157.
3   Ibid., p. 158.

The Semantic Web is a reality best represented by search engines, which are programmed to return results based not only on what we've requested in the search string but also considering the kinds of things we've searched for in the past on the same computer and things other people are searching for.

The best example of the Semantic Web at work is Google-as-search-engine, which is so dominant that it's frequently verbed, as in "Google it."

Google's algorithms note patterns in your digital content use and focus not just on the search term but what *you* are likely to be looking for.

Take the word "apple," for example, as a term you've typed into a search string. If you recently searched for things like crop insurance, weather reports, and antifungal treatments for trees, Google will return information sources relating to the fruit. If you recently searched for things like tech gear and new devices, Google will return resources relating to Apple, the consumer electronics company.[4]

Google is programmed to optimize user experience by tracing your path across the web via the range of digital devices and apps you use. An added benefit for Google-owning Alphabet, which has a considerable stake in digital advertising, is the ability to target and sell your interests as it tracks the little pieces of you that emerge through searches, social media views, follows and likes, and web browsing.

A Reminder: The Web Is Not the Internet

While the web completely relies on the physical structure of the internet to function, it largely comprises content we've collectively created.

The web is markedly different, by design. It effectively is a rhetorical platform—a space created to enhance communication and collaboration—that thrives because *we* contribute virtual pieces of our selves.

It's difficult to come up with a clear analogy of what that *looks* like.

One way is to think about the relationship among the internet, web, platforms, and content as a potluck, where everyone is invited to bring a favorite casserole, side, salad, drink, or dessert to share with others.

The internet, in this imagining, is like the room where the potluck is being held, which may have an endlessly changeable quantity of things inside of it.[5] The web, then, is one of many things available in the room, like the table all of that potluck bounty gets piled upon.

The various bowls, pans, and platters added to the table are akin to the role of platform service providers—divisions that define and separate different types of content. The range of content within those containers are the consumables, the food and drink individuals share to the potluck.

The table can be removed from the room and the room will still exist. But taking away the table effectively disrupts the potluck—the buffet of contained content has no common place from which people may fill their plates.

In short, the web is a relatively small part of the internet and relies on the internet to function. The internet is a relatively stable, malleable network with or without the web.

Also, you can't get to the World Wide Web without going on the internet, regardless of how you're trying to access it. Think about the room above as having doors and windows. You technically could enter the room through either, but you do have to enter that space to get to the table.

That access, whether on traditional computer or mobile device, is through a **web browser**, which effectively interprets the coding that makes up the web and allows you to move through it seamlessly.

---

4   Unless, of course, you are searching for an Apple product that directly competes with Alphabet's Android operating system. In which case you may also get Android or Google partner product results, just because Alphabet can.

5   Take a quick look at the room around you: How much stuff is in it? Think about the space among the stuff—floor space, air—there's a lot of space that could be filled, even in a very small room.

With smartphones, your default browser may be a **native app**, a mobile application that came pre-loaded on your phone. For example, Safari is the native web browser app for iPhones, and Samsung phones come with a proprietary web browser app that is simply labeled as "Internet." You also can choose to download apps for Chrome, Firefox, or a handful of alternatives, with all allowing you to access web content.

Apps like Facebook, Instagram, and Twitter include what's known as an **in-app browser,** which keeps you within the app but allows you to open web content without leaving it. This is what allows you to click through to shopping sites and news articles from the social media platforms, which effectively opens a new window within the app for the content.

Apps also don't run on the World Wide Web. Effectively the downloaded app runs on your phone and is updated frequently via internet connection, which allows some apps, like games, to function when you aren't online. That points to a key difference between an app and a web browser: You can't open a website if you're not connected to the internet.

Most social media sites allow at least some functionality, if not full participation, via the web and apps, from desktop, to laptop, to tablet, to smartphone.

Regardless of how you get there, the purpose of sites like Facebook, Instagram, LinkedIn, Pinterest, Reddit, Snapchat, TikTok, Twitter, and YouTube is to host **user-generated content,** the individual contributions of stories, images, videos, and random musings that are the core of today's World Wide Web.

Without our individual contributions—if we only consumed, rather than posting—the feeds would stop and these companies eventually would fail. YouTube would become the online equivalent of TV Land, reruns *ad infinitum;* TikTok would be an archive akin to "America's Funniest Home Videos"; Facebook pages would become little more than static digital scrapbooks; Twitter would simply vanish.

This is an increasingly complicated system built for participation and collaboration, which is exactly what architects of the World Wide Web had in mind. But as the web has evolved, it's also deviated markedly from the egalitarian, community-based world people like Tim Berners-Lee imagined.

Technology Is *Supposed* to Evolve

Berners-Lee was clear that the web always would be a work in progress—construction that never ends—and that it would take a number of evolutionary steps to approach the technology's potential. Berners-Lee outlined four steps to achieving his dream. Some already are commonplace, while others await further technological advancement:

1. **Affordability**

    This isn't about individual ability to buy digital devices or software. Computers, including smartphones, are hardware through which the web flows, much as television is hardware through which entertainment and news flow. Not owning a TV doesn't mean you can't watch TV. Not owning a computer doesn't mean you can't participate on the web.

    Hardware and connectivity issues are resolved in a range of ways, through things like public access computer labs (the new role of libraries) and free wireless access (the new role of Starbucks, and a significant security risk if you're not careful).

    Affordability, from Berners-Lee's perspective, is about information access and the ability to contribute original content. People should be able to get information and basic accessibility tools—browsers, search engines, programs, applications—for free and then upgrade as they choose.

    You can choose to pay for digital news subscriptions, or not, without stopping the free flow of information. You can download a free web authoring program and build a web page from scratch. You can listen to streaming music, with a set number of skips and ads every few songs. You can have a free LinkedIn account to promote yourself professionally.

But you also can spend hundreds of dollars for a robust web authoring program that lets you do far more than the free program; you can pay a small amount to subscribe to a music platform, choose your song order, and avoid ads; you can pay for LinkedIn Premium and see who's checking out your profile.

2. **Permanent access**

Whenever you want to access the web to share information or an idea, you should be able to do it. *Right now.* No waiting.

Berners-Lee was particularly vexed by dial-up, which took far too long to connect to the internet. This infrastructure issue has and hasn't been solved (see Chapters 9 and 12).

Today's fiber-optic, broadband, cell network, and wireless services may be more expensive, but they are far faster. And smartphones and tablets certainly allow us at-our-fingertips access to the web.

3. **Location independence**

This is about having access to your files independent of where the data are stored. Berners-Lee didn't think the web should be relegated to a window on a desktop computer, with all your personal settings stored *there*. The solution comes down to a now-ubiquitous term: cloud computing.

That's is simply the idea that user-generated content doesn't have to be stored on a single home- or office-based computer, where that data is vulnerable to deletion or damage; files can be backed up—often automatically—to a remote server and accessed from any digital device via the internet. (More on the complexities of cloud computing at the end of the chapter.)

4. **Protocol independence**

This is where things get complicated. Despite the elegance of web browsers, programs, and apps, we still don't have a sole intuitive interface for creating, editing, uploading, and downloading information. And maybe we really don't want one.

High reliance on solo technologies can result in havoc wreaked by things like hacker attacks and viruses. Those prone to technological mischief love it when companies connect all products via a single protocol. Makes it much easier to hack. For example, in 2015 hackers discovered an exploit in the Uconnect system built into Chrysler, Dodge, Fiat, Jeep, and Ram vehicles that could theoretically allow all such-equipped vehicles to be remotely stopped at once.[6]

As you learned in Chapter 9, there are myriad protocols that must be engaged for communication between your computer, the web, and other subnetworks on the Internet—TCP/IP, HTTP, POP, SMTP, SFTP, and so on.

Say you want to craft your own website using a web authoring program. You create your page, save, upload, view. If only it were that easy. There are myriad little steps that must be executed with care.

Today we are hindered by mishmash of protocols and programs that have to be carefully linked together, in the right order, to post professional-looking original content. Sometimes things don't work and sometimes—almost inexplicably—they do.

There are a large number of service providers on the web that allow us to create content without engaging protocol but only within their stifling parameters (log-ins, user agreements, and

---

6   "Hackers remotely kill a Jeep on the highway—with me in it" by Andy Greenburg, Wired.

limited, standardized options, to start). This resolves some protocol issues but generates other problems that undermine the founding ideals of the web.

Contemporary platform service providers—social media, in particular—hold us captive within proprietary, intuitive interfaces that shield both the complexity of the system and its ability to invade our privacy. This reduces our understanding of the technology, which ultimately may stifle innovation.

Challenges to Achieving Tim's Dream

The web was designed for collaboration, and collaboration requires **open systems**—those that encourage wide participation and nurture diverse viewpoints. But there are conflicting ideologies of what the web *is* and what it could/should be.

The long dominance of legacy news media in the United States sometimes is presented as a tyranny of experts who told us what to think about by choosing what issues would be covered and framing stories in particular ways. It was a **closed system** that didn't allow people who weren't journalists to contribute or collaborate. But closed systems were an invention that reflected the technology on which they were built (i.e., paper).

Then, starting in the mid-1980s, people gained access to home computing. The notion of platform at that point was that a computer functioned through a particular operating system (like PC DOS or Mac OS) that excluded applications not designed for it.

Users could choose an **Online Service Provider,** like CompuServe or AOL, to access the internet. These were closed systems, but they allowed people to interact freely behind a subscription paywall.

By the late-1990s, the World Wide Web was offering free access to an astounding amount of information, which quickly eroded the Online Service Provider subscription base. Users needed only to choose an **Internet Service Provider,** which connected an individual computer to the Internet via telephone lines or cable television, and then use a browser (like Microsoft Explorer, Netscape, or Opera) to connect to the web.

The early 2000s saw gains to open systems, including the rise of crowdsourced databases like Wikipedia. At last there was a framework that allowed people to share their own expertise on particular topics and consume information shared by others.

From about 2007 forward, however, the wide-open nature of the web has been challenged by an increasing number of **controlled systems**.

Platform Service Providers—companies that provide you space to participate, like social media—require users to sign up, explicitly agree to user terms, and log in if they want to participate. In exchange, users get personal space for which they can regulate other people's access.

Controlled systems take on many of the characteristics of a gated community in the physical world:

–   Access to the system is guarded by privacy settings, permissions, and sign-ins;

–   Access to individual pages is granted only by user approval;

–   Access gets even more exclusive, in that we can exclude those whose beliefs may challenge our own. We also can exclude negative nellies and people who share all-to-real personal problems. (Buzz kill!)

–   There's a cost to all this exclusivity. In the physical world we pay with our dollars. In the virtual world we pay with our privacy. (See Chapter 15 for more extensive discussion of privacy.)

The smartphone revolution brought us mobile applications, or apps, many of which function as **walled gardens** that confine our activity to within the app. This marks a significant return to closed systems.

Apps comprise their own little closed universe, as the user only has to click on the app icon to gain entry, rather than opening a browser and going to a website. In fact, if you're looking at something on a mobile device, you probably got there via an app—email, Skype, iTunes, Facebook, LinkedIn, Pinterest, Instagram, Snapchat, digital radio, maps and navigation, banking, fitness, etc., etc., etc.

As previously noted, in-app browsers allow users to click on links that take them to web content, but they do so without ever leaving the app. When you click on "Shop Now" on an ad in Instagram, for example, the shop's site opens on the web. The little X in the top left-hand corner allows you to close the window and return to Instagram—which targeted your personal interests for the ad in the first place.

The intent of many apps is to keep the user in place or allow movement only within a narrowly defined system. The ever-popular *Candy Crush Saga,* for example, provides a variety of rewards to the user for staying in game developer King's controlled system. It also makes recommendations for other games created by King (e.g., *Bubble Witch* and *Candy Crush Soda*), which keeps the user within King-controlled space.[7]

The more popular an application is, the more attention it pulls away from the web. And eyeballs and web clicks are what make money for Platform Service Providers.

Part of the attractiveness of apps is the *perception* that they don't pose the same threats as the web for things people increasingly want to avoid: privacy, corporate colonization, hyper-commercialization, identity theft, government surveillance, viruses, and other unpleasantries such as denial-of-service attacks launched by hacker groups.

But closed systems, like mobile apps, have the capacity to isolate us from one another and disrupt the potential of the World Wide Web. They allow us to ignore what we don't want to see or hear, which doesn't come close to meeting a goal of distributing knowledge across time and space *for all.*

How Do We Save the Web (and Save Ourselves)?

It's arguably accurate to say that a person has reached many markers of adulthood by the age of 30. Maturity and a level of common sense is likely to have set in. For the World Wide Web, that milestone birthday felt a little more like rebellious teen angst.

Sir Tim Berners-Lee, who was knighted by Queen Elizabeth in 2004 for his invention of the web, marked that birthday by praising the web's influence and lamenting its misuse:

> The web has become a public square, a library, a doctor's office, a shop, a school, a design studio, an office, a cinema, a bank, and so much more.[8]

> I had hoped that 30 years from its creation, we would be using the web foremost for the purpose of serving humanity … However, the reality is much more complex. Communities are being ripped apart as prejudice, hate and disinformation are peddled online. Scammers use the web to steal identities, stalkers use it to harass and intimidate their victims, and bad actors subvert democracy using clever digital tactics.[9]

---

7   While King may be the name on the game, the company was acquired by Activision Blizzard for $5.9 billion in 2016, with that company acquired by Microsoft for $68.7 billion in 2022.
8   "30 years on, what's next #for the web?" by Tim Berners-Lee, writing for the World Wide Web Foundation.
9   "I invented the World Wide Web. Here's how we can fix it" by Tim Berners-Lee, The New York Times.

Placing blame squarely on governments, social networks, and users, Berners-Lee noted three particular problems that endangered the future of the web and civil society.[10] He also offered solutions:

1. **Deliberate, malicious intent**, such as state-sponsored hacking and attacks, criminal behavior, and online harassment.
   Solution: Calling on governments to update laws and regulations to meet the moment, as well as to protect an open web, rather than allowing companies to sacrifice individual privacy and democracy for grotesque profit.

2. **System design that creates perverse incentives** where user value is sacrificed, such as ad-based revenue models that commercially reward clickbait and the viral spread of misinformation.
   Solution: Embracing digital design that doesn't sacrifice privacy or security, as well as encouraging those in the tech sector to continue exposing profitable corporate practices that "come at the expense of human rights, democracy, scientific fact or public safety."

3. **Unintended negative consequences** of benevolent design, such as the outraged and polarized tone and quality of online discourse.
   Solution: Urging web users to take personal responsibility for fostering constructive, healthy conversations online; empowering them to hold governments and companies accountable for systemic issues; and demanding intervention in what's become an unhealthy, unsustainable digital reality.

Berners-Lee pointed to the web's issues 30 years in as part of "our journey from digital adolescence to a more mature, responsible and inclusive future."[11] He cautioned against acceptance of domination by companies that use our own data to surveil and sell us, making enough money along the way to stifle innovation.

> We're at a tipping point. How we respond to this abuse will determine whether the web lives up to its potential as a global force for good or leads us into a digital dystopia. [12]

While Berners-Lee wants governments, companies, and individuals to take responsibility for the state of contemporary technology, he isn't waiting for someone else to lead. He's working on an open-source start-up called Solid-Inrupt, which he thinks could return personal data sovereignty and trust in societal institutions. Much as with early development of the World Wide Web, most people can't quite wrap their minds around what Berners-Lee is trying to do, other than, as one researcher noted, "he's on the right side of history."[13]

It's not hard to imagine that the invention of something as influential as the World Wide Web today would be purchased and monetized, or possibly killed as competition, by any one of the Frightful Five. Somewhat ironically, that would have cost Berners-Lee the opportunity for knighthood, but might actually have made him the billionaire he's not. His net worth in 2022 was $10 million, while Mark Zuckerberg, one of many who built an empire on the web Berners-Lee wove, had a net worth of more than $65 *billion*.

10   "30 years on, what's next #for the web?" op. cit.
11   Ibid.
12   "I invented the World Wide Web. Here's how we can fix it." op. cit.
13   "He created the web. Now he's out to remake the digital world" by Steve Lohr, *The New York Times*.

On the Web or Not, Where's My Stuff?

Cloud computing allows us to access files from any computer, but it's not something that people tend to actively manage well for two basic reasons:

1. **Poor file management.** If you're one of those people whose virtual desktop is cluttered with dozens (if not hundreds) of individual files, or if everything goes into your computer's Documents folder, you need to learn File Management 101.[14] If you can't find a file on your computer, good luck finding it once it's in the cloud.
2. **We generate a lot of content.** You're pretty clear that if you drop your laptop it dies and takes your files along with it. But at least you knew where they were to start with.

Most computers, programs, and apps have an option to automatically save to a distant server—the cloud—rather than to your computer, so you always have a backup. When you think about the potential for catastrophic damage to your phone, say from dropping it in a toilet, the comfort that your content is automatically backed up is awesome.

But what happens when our phone contacts, calendar, photo gallery, and text messages have been archived by Verizon and we change to Google Fi? If it's not *our* cloud, can we still access and/or transfer the content?[15]

You may work within multiple cloud services a day, many retrieving our content automatically, so how do you keep track of where all your information is stored? And what about content you *really* don't want to be anywhere but on your phone? Can others access it once it's in the cloud?

The latter question has been answered repeatedly by high-profile hacks of cloud servers. One of the most notable took place when a hacker exploited security flaws in iCloud and Gmail servers to harvest hundreds of celebrity nude selfies. In what was later dubbed "Celebgate," images of Jennifer Lawrence, Kate Upton, Ariana Grande, and others were stolen and posted online in 2014. Three hackers were convicted of violating the Computer Fraud and Abuse Act and sentenced to 6–18 months in prison, but there was no stopping the spread of the images once they were online.

I Get That I'm Saving to It, But Where's This Cloud?

Our enthusiasm for "the cloud," which today comprises millions of remote servers, has clear implications for internet security, digital living, and our physical world. This consolidation of information into massive data centers poses a range of concerns, from energy consumption to privacy to potential for sabotage and cyberattacks.

While the vast cable system and redundancies of the internet provide solid infrastructure, the bulk of the information—the content you upload and download—is stored in and accessed from **data centers** across the United States. These centers power, protect, and maintain millions of networked computer servers, stacked floor-to-ceiling in expansive, climate-controlled spaces.

There are close to 100 data centers[16] (aka the cloud) in the United States, but the exact number of either centers or servers is difficult to find, as the companies tend to be somewhat secretive about their

---

14 Here's the basic idea: Folders. Folders. Folders. Technically it's known as hierarchical organization, which is just a two-word way of saying, "put everything that goes together within one folder, and sort related files into subfolders." Check out Appendix 5 for greater exploration of file management.
15 Answers may vary. Check before changing services.
16 At this writing, public information online indicates that Amazon Web Services is the largest cloud services provider with at least 30 data centers in the United States. By comparison, Alphabet has 14, Apple 6, Meta 17, and Microsoft 12, with more under construction. Each company also owns servers elsewhere in the world.

storage capacity and data center locations. And with good reason: Destruction of one data center by either natural disaster or human mischief would result in a great loss of content.

Some data centers are solely owned by Alphabet, Amazon, Apple, Meta, and Microsoft, while others lease space to thousands of web service providers. Apple also pays Alphabet and Amazon millions of dollars each month to store iCloud data on their servers.

Many data centers are in rural areas, taking advantage of greater availability of and lower prices for land. Oregon is home to a few data centers that take up space in places like Prineville, Umatilla, and The Dalles, which have a combined population of fewer than 30,000 people. Data centers typically employ fewer than 100 workers and generally have off-the-grid backup systems fueled by diesel engines or solar power.

Data centers tend to take up a lot of space, as well as consuming a stunning amount of energy. Many data centers in the U.S. are close to 300,000 square feet. To put this in perspective, a typical Walmart Supercenter is smaller than 200,000 square feet, a World Cup soccer pitch is just shy of 70,000 square feet, and a regulation NFL football field is 57,600 feet. The *smallest* commercial data center is only slightly larger than a combination of all three and 48 football fields could fit inside the largest. Wrap your mind around *that*.

Figure 10.2. The relative size of a 300,000-square-foot data center.

The largest data center in the U.S., The Citadel Campus, has 7.8 million square feet of server space—that would be *39* Walmart Supercenters or *111* soccer pitches—with room to grow to 17 million square feet. That data center is located near Reno, Nevada, right next door to the Tesla Gigafactory, which produces Tesla batteries. The Citadel Campus is owned by Switch, a company based in Las Vegas, Nevada, that specializes in server hosting.

Data centers are home to millions of servers, individual hard drives, and are underpinned by hardwired technology that, just like *your* technology, is powered by electricity.

We're Not Saving the Planet

Paper, the kind that makes books and newspapers, is a renewable resource. Trees can be planted, grown, and harvested. Paper is recyclable and degrades rather quickly when exposed to the elements. Paper manufacturing plants and the paper recycling process, however, are notorious for energy consumption and water pollution. That said, there are far fewer paper plants today than there are data centers. Is that progress? Let's find out.

Data center servers don't sleep the same way your computer does if you leave it idle for a while. They are part of an on-demand system that feeds us instant answers to important and trivial questions

24/7. Much as your computer burns battery when it's being used, data center servers burn electricity as they wait for your call.

Servers also generate a lot of heat and data centers must find efficient ways to cool the air around them, most of which utilize a lot of water. For that reason, places like the Columbia River Gorge, which generates both hydroelectric energy and water for cooling, are an ideal location.

To improve data speed, centers need to be scattered in various regions, so their locations are based on consumer convenience, rather than climate realities.

In 2020, Alphabet received approval to build a data center near Mesa, Arizona, an area of the country that is arid and hot, at best. In exchange, Alphabet selected Mesa to be the first Arizona city to have Google Fiber—low-cost, high-speed internet.[17]

As an incentive for locating the $1 billion data center in Mesa, Alphabet was guaranteed up to 4 million gallons of water per day.[18] To put that amount in perspective, the Environmental Protection Agency says that each American uses an average of 82 gallons of water each day.[19] Alphabet's maximum daily water allotment is the amount of water consumed by nearly 50,000 people.

About a year later, Meta broke ground on a massive data center complex in Mesa, approved for up to 1.7 million gallons of water each day. Rather than drawing from the Colorado River, as other area data centers were doing, Meta promised to restore more water to the area than it would use as it worked to be "water positive" for all of its global operations.

A year later, Arizona entered its 22nd year of drought and Colorado River reservoirs, from which 40 million people across seven states draw their water supply, hit a record low. The general manager of the Colorado River District pointed to decades of overconfidence in a dwindling water supply as part of the problem: "We thought we could engineer nature. Huge mistake."[20]

These are just two examples of choices being made by communities and corporations every day to meet our collective demand for everything, right now. More data means more data centers, which require more land, energy, and water.

While contemporary data centers are increasingly energy efficient and strive for environmental sustainability, the fact is that billions of digital devices, pulling information from and sending it to millions of computer servers, do not make for a neutral environmental footprint.

It's an on-demand world and on-demand comes at a cost.

Summary

The World Wide Web is a communication platform that we've collectively created, using a variety of programs and technologies that allow varying degrees of user interaction. Web architect Tim Berners-Lee envisioned a space where people could collaborate (intercreativity), with the resulting content organized and analyzed by computer programs (the Semantic Web). Berners-Lee imagined that the web would evolve as technologies evolved but ultimately function best as an open system of information. Highly controlled systems, such as social media platforms, and walled gardens, such as smartphone apps, pose a challenge to Berners-Lee's vision of the web as collaborative space.

17 " 'Google Fiber is built for speed': Mesa to be next city to receive the high-speed internet service" by Corina Vanek, Arizona Republic.
18 "The secret cost of Google's Data Centers: Billions of gallons of water to cool servers" by Nikitha Sattiraju, Time.
19 From the EPA's WaterSense Statistics and Facts: epa.gov/watersense/statistics-and-facts.
20 "The West's most important water supply is drying up" by Conrad Swanson, The Denver Post.

Key concepts

Closed systems
Cloud computing
Controlled systems
Crowdsourcing
Data centers
In-app browser
Intercreativity
Internet Service Provider
Native app
Online Service Provider
Open systems
Semantic Web
User-generated content
Walled gardens
Web browser

References

Berners-Lee, T. (2019, March 12). 30 years on, what's next #for the web? *World Wide Web Foundation*. Retrieved from https://webfoundation.org/2019/03/web-birthday-30/

Berners-Lee, T. (2019, November 24). I invented the World Wide Web. Here's how we can fix it. *The New York Times*. Retrieved from https://www.nytimes.com/2019/11/24/opinion/world-wide-web.html

Berners-Lee, T. (2000). *Weaving the web: The original design and ultimate destiny of the world wide web by its inventor.* New York, NY: HarperCollins.

Greenberg, A. (2015, July 21). Hackers remotely kill a Jeep on the highway—with me in it. *Wired*. Retrieved from https://www.wired.com/2015/07/hackers-remotely-kill-jeep-highway/

Lohr, S. (2021, January 10). He created the web. Now he's out to remake the digital world. *The New York Times*. Retrieved from https://www.nytimes.com/2021/01/10/technology/tim-berners-lee-privacy-internet.html

Sattiraju, N. (2020, April 2). The secret cost of Google's data centers: Billions of gallons of water to cool servers. *Time*. Retrieved from https://time.com/5814276/google-data-centers-water/

Swanson, C. (2022, July 21). The West's most important water supply is drying up. *The Denver Post*. Retrieved from https://www.denverpost.com/2022/07/21/colorado-river-drought-water-crisis-west/

Vanek, C. (2022, July 1). "Google Fiber is built for speed": Mesa to be next city to receive the high-speed internet service. *Arizona Republic*. Retrieved from https://www.azcentral.com/story/news/local/mesa/2022/07/01/mesa-set-1st-arizona-city-receive-google-fiber/7778615001/

CHAPTER 11

From Neighbors to Followers: Rethinking What It Means to Be Part of a Community

Figure 11.1. *The notion of selfies—taking pictures of ourselves to post on social media—didn't take cultural hold until smartphones proliferated, apps became ubiquitous, and forward-facing cameras improved (circa 2013). Pearls Before Swine © 2017 Stephan Pastis. Reprinted by permission of Andrews McMeel Syndication. All rights reserved.*

Takeaway: Physical and digital communities are what we make them. They can complement or destroy each other.

People, for the most part, have a basic need for each other—to talk, to be heard, to touch, and to be touched. But we can never truly know another person. What we actually *know* is only a part of any person, as we each have multiple and conflicting identities that exist in the complete isolation that is our physical bodies.

We can talk to people, we can listen to people, we can be around as many people as is people-y[1] possible—but at the end of every interaction, we're back to being alone with the voices inside our heads (the normal voices, not the other ones).

And we can't always choose who we spend large chunks of time with in the physical world. We're born into families, assigned to classrooms, moved into neighborhoods, hired for jobs. *Not. In. Your. Control.* We may not even like people we have to interact with frequently, but most of the time it doesn't matter; we're all in this life together.

A physical community is a lot like a prickle of porcupines huddled for warmth under a log on a cold winter's night: We can't help but poke each other, but we need each other to survive.

How we decide with whom to huddle[2] has been the source of an endless range of studies within numerous fields of study. Theorist Kenneth Burke[3] said there are three basic ways we interact with each other and our environments:

1. **Identification by Sympathy**—We have something we *like* in common.
   Physical world example: A group is choosing films to stream from an online service. The people know each other, but it doesn't really matter how well. One person says, "Let's look at the horror flicks—there's a great selection." Another person nods enthusiastically, saying, "Let's start with 'The Ring' and see how many we can watch. I love horror movies!" The group settles in, united in common purpose—the desire to binge-watch horror movies.
   Virtual world example: Who we choose to friend and follow on social media, as well as what kinds of views we expect to show up in our feeds. What happens when those views are contradicted?

2. **Identification by Antithesis**—We have something we *dislike* in common.
   Physical world example: While the ultimate example is war, let's continue with the less complicated example started above: The person next to you says quietly, "I hate horror movies." You give a relieved smile and say, "So do I. Let's go in the other room and stream some rom-coms. I'm in the mood for funny." A couple of other people follow, drawn largely by the desire not to watch horror movies.
   Virtual world example: Friending and following people on social media who are opposed to the same things you are.

---

1   People-y is a made-up word, but it is preferable to and more fun to write than humanly.
2   And, no, that's not a euphemism for something else. We're still talking about people as porcupines. Stay with us here …
3   Burke explored human nature, particularly motives for communication and behavior, throughout his career. Identification is most notably covered in his book, *A Rhetoric of Motives* (1969).

3. **Identification through Transcendence**—Identifying with something larger than ourselves gives us an intense feeling of personal freedom. Burke termed this "the principle of the oxymoron,"[4] or *misidentification.*

   Physical world example: You stand on a mountaintop and feel the mountain's power. You essentially are transcending your own limitations by identifying with the mountain's greatness, which is not and cannot be your own.

   Virtual world example: Posting an image on Instagram and having it generate several hundred likes can provide a similar feeling of transcendence—just by virtue of posting, you may mistake the image's power—and that of the medium—for your own. But you are not the internet or World Wide Web; you barely register as a cog in a vast technology machine.

Each form of identification applies not only to our physical lives but also to our digital lives. They are the things that result in community, whether in physical space or cyberspace.

It's important to note that people don't actually have to have anything in common with others, or even interact with them in any meaningful way, to identify with them. That's because we form *perceptions* of where those interests meet and diverge—the identification of one person with the ideals of another does not have to be shared or reciprocated to be a motivating force.

People's identifications are situated in context, the result of both a lack of choice—we can't change the life we're born into—and things over which we have more influence, like lifestyle and livelihood. The result is a sort of symbolic blinder that prevents people from seeing or understanding the lives of others.

Take, for example, an aging farmer, raised in rural Idaho, who never attended school past the eighth grade. Now put that person in the room with a teenage social media influencer from Connecticut. Their views of reality will be very different and their ability to communicate in any meaningful way will be limited. Each may consider the other to be ridiculous, while both possess significant knowledge and abilities that allow them to survive and even thrive in the world they inhabit.

Many television shows and films are based on such a "fish out of water" premise that brings together people from different places and social groups, often with comic results. Such deliberate *mis*identifications point to both deeply imbedded social division and the endless quest for common ground.

While initial focus might be on how the aforementioned pair are different, forcing them to remain in a room together will momentarily link their fates. It is likely, in the Burkean worldview, that they would ultimately begin the verbal quest for some sort of identification—whether it be in terms of things they hold in common, those they mutually disdain, or the symbolic achievements each can point to as evidence of their personal value.

While physical space dominated notions of community for thousands of years, along came the internet and personal computers and we suddenly could escape the confines of our physical lives and *be* whoever, *with* whomever, wherever, and whenever we chose. Voilà! Online community, mediated by a range of digital technologies, was born.

---

4   Burke, op. cit., p. 324. Burke is great to read, by the way: a little dense, but interesting, just in case you want to go to grad school or something.

Digital Communities Are a Lot Like Physical Communities

This brings up a key question: Are online communities *real* communities?

Is an infinite flood of information that we each access from a different physical space, staring into individual screens, conducive to the level of communication required for community building?

Does digital communication effectively reproduce and replace human connection?

If finding **common ground**—the places we perceive our interests as intersecting with others—is how we begin to identify with one another, does technology help or hinder the level of communication necessary for enduring connection?

On one hand, the convenience and ubiquity of digital communication technologies allows us to find each other and build meaningful connections around topics of real interest.

On the other, there's an inherent divisiveness to digital communication driven entirely by individual preference.

A fundamental benefit of mass communication was that it gave people common, shared signals that served to link and maintain greater society. As recently as the early 2000s, it could have been claimed that the United States possessed the semblance of shared culture and identity, facilitated largely by the mass media. One need only look to the cultural touchstones embodied by Disney, shopping malls, television shows, big box stores, and franchise restaurants to understand the influence of shared mass culture. While it's easy to look fondly on what seemed like kinder, gentler times, we also must recognize that the American culture and identity facilitated by mass media was one created by some to the exclusion of others.

Part of the early and ongoing fascination with digital communities is the perception that they are a space where everyone can find a place to belong, a place of common ground. A utopian view would describe these communities as:

- Nonjudgmental

- Welcoming

- Uncritical

- Places where appearance is not an issue

That certainly explains why digital communities—from support groups, to gaming, to social media—are so popular. But are those assumptions true? Not entirely. Active participants, depending on the platform, are more likely to be:

- Judgmental

- Territorial and have a member hierarchy based on time invested in participation

- Skeptical, hyper-critical

- Spaces where appearance matters greatly

Odds are this isn't a surprise. Humans are prone to pack behavior, and digital communities, particularly social media platforms, are literally designed for it. We join, follow, add, connect, comment, and share in an effort to validate our perspectives by aligning them with others. Consider here the phenomenon of Twitter dogpiling, where seemingly everyone piles on criticism of a thoughtless tweet from a stranger in a mass act of public shaming.

And think for a minute about your own online representation. The images you select to represent yourself on various sites are called **avatars** and are part of who you are in the digital world.

People who participate in digital communities tend to take great care with their avatars. Think about your various social profile pictures. Why don't you leave the platform placeholder? Does it matter how you look? How often do you change it?

The advantage/disadvantage of digital is that we can separate our physical selves from our cyber-selves. We can, in effect, have multiple identities, occupying space in multiple communities.

There's an interesting piece to consider here, the concept of the divided self and what it means to be part of multiple communities. Where are our loyalties? To our physical world? To *one* game, app, or social media platform? Our communal selves are fragmented by the demands of the physical and desires of the digital world

It's critical to recognize that our physiological needs for things like food and comfort, coupled with our psychological needs for emotional and intellectual support, and tendencies toward identification, can lead us mistakenly to adopt and adapt to the perspectival concerns of others. It is a warning to heed: We are individuals and are distinct from other people. What we see is what other people knowingly or unknowingly allow us to see. It is not so difficult for our needs and desires to be recognized, seized upon, and manipulated by others. Particularly in the digital world.

So, yes, technological change has made significant adjustments in our perceptions of the world—some good, some not so. But, as Kenneth Burke famously noted, "an 'adjustment' need not be a 'good adjustment.' It is as much an 'adjustment,' in our sense, for the organism to die of a bullet as for the organism to dodge one."[5]

Traditional Versus Digital: Communities in Transition

The definition of community has been in flux since the mid-1950s, due largely to the confluence of air conditioning and television, followed by a host of other inventions that lured us from our porches, backyards, neighborhoods, theaters, and city centers, to our couches and other personal spaces.

Consider the telephone. Not the one in your pocket, but the back-in-the-day, stuck-to-the-wall-of-your-house version. They had a significant role in changing community and community building. Before phones, people living within one community might be inclined to meet in person—to visit. After phones, there was the convenience of near-instant connection.

For a long time, neighbors, particularly those living in rural areas, shared telephone service lines. They were called "party lines" and when you dialed a number it would ring at all houses hooked up to the common phone line.

That meant when you made a call you never knew for sure who would answer or who else would be listening. You ring your mom and you end up talking to her neighbor, then grandmother, then brother. Finally, the phone is passed to your mom, but you've gained a good deal of mediated human interaction along the way.

Later, most houses had single phone lines and, often, multiple phones. But still, all you had to do was to pick up a handset to hear—and perhaps participate in—someone else's conversation.

Our cultural shift to mobile phones means that fewer households have landlines and party lines have ceased to exist. We have individual phones and phone service that are meant for our communication, participation, and social engagement *alone*. When you want to talk to someone, you call that

5   Burke (1968), p. 151.

person's number. If the person isn't available, you can leave a voice message, or text, or direct message them via social media. Odds are that no one but that person will answer, hear, or respond to your messages.

This would imply that we've gained a degree of communication privacy unheard of in past decades. But that's not the way it works. We tend to answer our phones and respond to messages wherever we are—restaurant, grocery store, library, doctor's office, bathroom stall. And, somewhat bizarrely (when you think about it), people on smartphones tend to share one-sided voice conversations with everyone nearby, acting as if they're in a personal Cone of Silence.[6]

As this example illustrates, physical and digital communities exist together, but are remarkably different things. What follows is a list of **community characteristics** in which physical and digital needs, goals, and purposes are significantly different.

Boundaries
- Traditional communities are *communities of space* and members must be in physical proximity to one another.

  A traditional community has physical boundaries that can be crossed, such as streets, neighborhoods, city limits, and county and state lines. Most communication is face-to-face.

  A classroom, for example, is a traditional community. There's a designated time for you to be in a particular space and, for the duration of the course, the community is active. There's an expectation of participation, even if that just means showing up.
- Digital communities are *communities of spirit.* These communities are formed around ideas and hobbies, rather than proximal necessity (like neighborhoods). They are flexible, effectively shrinking space and time by allowing members to join in whenever they are able. Communication typically is asynchronous, which means that you can contact someone on your time frame and they will answer on theirs. Most communication is via bulletin boards or instant messaging.

Economy
- Traditional communities once relied largely on the health of the *local* economy, which was strengthened by the norm of reciprocity—you shop with me and I'll shop with you. This is what those "Shop Local: Keep Our City Green" campaigns are trying to encourage—shopping locally to keep the local economy strong.

  This is one shift that is not due only to digital competition but also the increasing number of big box stores that employ local workers but filter profits out of the community. They often are criticized for not paying what's known as a "living wage," which means that people who work at these retailers generally cannot afford to shop locally.
- The explosion of digital shopping options—from eBay to Amazon to Alibaba—have facilitated growth of a *global* economy. Instead of buying books, shoes, or other needed stuff locally, you may choose to shop online first to get better prices. By doing so, you don't pay local taxes or support the local economy. You are, however, supporting an increasingly global marketplace.

---

6   The Cone of Silence was one of our favorite things about the TV show "Get Smart," which aired on TV from 1965–1970. The cone was a gadget aimed at keeping top-secret conversations top secret. It didn't work. Hilariously. "Get Smart" was made into a 2008 movie starring Steve Carell and Anne Hathaway. We recommend the original.

Institutions

- Traditional communities are rooted in *institutionalism*. Institutions are things like schools, churches, hospitals, and government—the focal places that guide, take care of, and govern our physical spaces.
- Digital communities are rooted in *individualism*. Traditional institutions typically have space in the digital realm, but their presence is diffuse and their influence diluted by the astronomical number of other places Web users can choose to be at any given time.

Socialization

- There's a funny term (funny ha-ha *and* funny weird) that techies have come up with for the physical world: *Meatspace.* And, yes, it's m-e-a-t, not m-e-e-t. Meatspace is where the meatbots hang out. Us. The human body that we rely on for life.
- Digital communities are the realm of *cyberspace,* which engages your eyes, ears, and central nervous system far more than the rest of your physical body.[7]

Control measures

- Our physical selves are bound to the *rules, regulations, laws, and codes* that govern a community. Most of us are quite aware of what would happen if you went into your local Safeway grocery store and sat down in the middle of an aisle to play your guitar a while. You'd be asked to leave. And if you refused, you'd likely be arrested.
- Digital communities are *self-policing,* which is accomplished by the corporations who own particular sites or users who serve as moderators. Your rights online are governed by user agreements.
  If you violate the terms of a site by, say, posting copyrighted material or harassing comments, you can be stopped from participating. The cyberpolice don't exist and the U.S. government is not going to stop you from doing much of anything online, as long as it doesn't cross into the criminal realm (child pornography or hacking, for example).

Information

- Traditional communities rely on *guided learning* and the *authority of experts*. Information tends to be presented in a filtered, preprioritized package, complete with importance cues that tell you when and where to pay attention. Newspapers are a good example of guided learning, as are traditional classrooms.
- Digital is a realm of *self-discovery* and *concealment of expertise*. You decide what's important to you at any given minute. You can choose to learn a great deal about topics you are interested in, as well as share your own expertise.
  Anonymity is something that levels the playing field online. A Harvard law professor and a high school dropout can both post information on a law page on Wikipedia. It's up to the community to decide what stays and goes.

Participation

- Traditional communities make decisions based on communal will, with participation *obligatory*. It's your democratic duty to vote, to show up, and to be a part of the decision-making process.

---

7 Cyberspace is a term that was coined by sci-fi author William Gibson in the early 1980s.

Your participation also tends to be public, as with demonstrations where picketers are clearly recognizable, or public meetings where speakers must identify themselves.

– With digital communities, participation is *voluntary,* not expected, yet the nexus of decision making may not be clear. To that extent, many digital communities have the qualities of a dictatorship, not a democracy.

Facebook, for example, is an online community made up of a billion little online communities. You can control—to some extent—what shows up in your feed and what other people see, but Facebook ultimately controls *all.* The company makes sweeping decisions without consulting its users. And, note, we said "users," not members.

Technology

– Traditional communities value technology for their ability to foster *growth and progress.* Issues such as illness, homelessness, addiction, teen pregnancies, and so on, however, present physical reminders that there are many problems that cannot be solved by technology.

– Infatuation with the latest digital wonder can border on ***technopoly,*** which is the deification of technology. Technopoly includes the perception that technological innovation and function are the result of some sort of god-like magic that renders our lives more meaningful.

Media consumption

– Traditional communities are rooted in the physical—information is transmitted via particular legacy technologies, like paper or television. Legacy media are characterized by ***exhaustive use.*** You can consume an entire newspaper, magazine, book, or television program—read every word, watch every moment—because the content is *static.* And it's the same every time you look at it. You may notice different things, but the text and plot stay the same.

Legacy media also are *linear;* they have a beginning, middle, and end, and they are supposed to be consumed in order. Think of binge-watching a TV show's entire first season on Netflix: Where do you start? At the beginning, of course. If you don't, you can't follow the story arc.

– Digital media are characterized by ***intensive use*** through brief interactions with individual apps, pages, and sites. It's simply impossible to consume the entire World Wide Web, or perhaps even a single website, because the content can be and is subject to continual change.

Digital media are *nonlinear;* information, by design, does not have a predetermined linear order. Do you want the definition for a particular word? You don't have to browse a dictionary. Instead, you put the term into a search string and get the definition—entirely out of context. And once you have what you need, you move on.

Risk to community

– Traditional communities are at risk from social and economic issues, but perhaps the most serious contemporary problem is that of ***benign neglect.*** This means that people go about their lives ignoring a situation, rather than assuming responsibility for managing or improving it. Think mental illness, homelessness, and lack of access to health care—pretty easy to ignore until it affects you.

– Typically online communities are maintained by a small group of people, with many others who come and go at whim. These people carry the burden of building and maintaining a community, which takes time and energy.

Online communities suffer when people lurk, rather than participate. A ***lurker*** is a Web user who looks at other people's posts without contributing to the conversation or committing to the community.

It's important to note that both traditional and digital communities are negatively impacted by audience preoccupation and fragmentation. When people's attention is pulled in multiple directions—whether in their physical or digital lives—participation in one, the other, or both suffers.

Now that you have some ways of thinking about old and new, physical and digital, imagine the future. Go on—close your eyes and think back to when you were 7. What did you think the future would look like?

Flying cars? Check. Space travel? Check. Time travel? Check. Robots? Check.

So, here you are now living in "the future" and what has changed? A lot … and nothing at all. That's because the future doesn't arrive all at once, obliterating the past; innovations emerge and social institutions adjust.

The future arrives a little at a time, bringing old ways and technologies with it. The past and future tend to coexist, which means that physical communities and digital communities are both seeking our undivided attention.

We note this here to clarify that, while the preceding section presents physical and digital communities as if they are distinct things, the characteristics of each cross over, blend together, and can be nearly indistinguishable from one another.

Following are two case studies of people who had dramatically different identities and community involvement in their physical and digital worlds.

Negotiating Identity: When Physical and Digital Lives Collide

Charlanne Corbin's century-old farmhouse in rural Magalia, California, caught fire shortly after midnight Tuesday, August 2, 2005. The 58-year-old died in the blaze, which spread to outbuildings and a neighboring field.

Corbin was a popular bus driver in nearby Paradise, California, fondly known as "Popcorn." This was a tragic loss. She had driven scores of children on her routes over the course of 21 years. Her supervisor noted that she was a "very popular driver," and coworkers said she was a good driver who inspired good behavior and "got along with the kids."

Yet Corbin was so much more than that. At the time of her death, she had been gaming online for 13 years—starting with the Internet and progressing to massive multiplayer games on the web.

Corbin was an active participant in *Battleground Europe,* a massive multiplayer role-playing game centered on the European battles of World War II. To those who knew her from gameplay, Corbin was not a bus driver. She was a French colonel by the name of Phew who commanded an axis battalion of like-minded gamers, the Pirates Battle Group.

While Corbin never met any of the other players in person, over the course of the years she became very good friends with many of them. And it was those people who contacted the Butte County Sheriff's Department to request investigation into her death, because they'd seen her online not long before the fire broke out. (The fire later was determined to have been caused by an electrical short.)

Her fellow gamers held a memorial service within the game the day she died and later had a larger service that included a cease-fire so all who knew Phew during her four years in the game could participate. They also added her name to the "missing in action" list posted in every town. Corbin also was named "Player of the Week" for the week she died and was the only player ever to be given the Medal of Honor in the game.

News of Corbin's death spread quickly online, with 132 posts made to the Legacy.com guestbook set up by *The Paradise Post* newspaper as part of its obituary service. Only 8 spoke to Corbin's life as Popcorn the bus driver, while 124—from gamers in 32 states and 12 countries—memorialized her as Phew.

A sampling of the posts demonstrates how different the identities of Popcorn and Phew actually were:

August 17, 2005
RIP popcorn you were one of my best bus drivers since kindergarden! [*sic*] my regards to her family! she always picked me up at nunneley and sawmill even though it wasnt [*sic*] a real bus stop **thanks popcorn for not making me late everyday and waiting for me to run to the bus!**

lv, Paradise, California

August 07, 2005
Charlanne and I became friends roughly 3–4 yrs ago. I knew her as Phew which was her game name in WWII Online. **I spent many of nights talking to her about everything under the sun.** She would help me when I needed advice about life issues. My deepest sympathies go out to her family and close friends around the globe. She touched so many hearts that its [*sic*] hard to fathom. Her love of people and for animals is what I will remember forever.

AK
Boston, Massachusetts

August 18, 2005
I first met Charlanne "Phew" online in mind [*sic*] 2001 while playing World War II online. I've played along side [*sic*] with her almost every day since then. **I never thought how hard it would be to lose an online friend. I find my self** [*sic*] **dealing with this as if I lost my best friend.**

QP,
Antigonish, Nova Scotia

August 08, 2005
Phew: **Most Folks don't understand our connections in the gaming world to most it is only a game!** But your kind voice and the funny things you said and did will forever be in my memory. **Today is the first time Ive** [*sic*] **seen your face** and I feel great sorrow for the people in your real time world!

GC
Marion, Indiana

August 08, 2005
I knew the 'virtual' phew. When I learned of her passing I was in shock and disbelief. **I never knew the death of a person I have never met would move me to tears.**

B
St. Albert, Alberta

Which was the real Charlanne Corbin—Popcorn or Phew—and what do these memorial posts indicate about a life lived via two communities?

Technoblade Never Dies

In 2005, Charlanne Corbin's connection to her online gaming community illustrated the deep relationships that can be built through MMORPG[8] play, which allows many people to play a game simultaneously online. Since Corbin's time, the technology underpinning the technology has increased and the number of games has multiplied. On any given day, games like *Final Fantasy*, *World of Warcraft*, *Fortnite*, and *Farmville* each attract millions of active players every day, cumulatively drawing hundreds of millions of sign ups.

---

8 Popular shorthand for massively multiplayer online role-playing game.

Games like *Minecraft* have evolved to occupy a hybrid world, with users able to play together online, as well as playing alone offline. Some users play alone, but film themselves playing and can draw a large fanbase by livestreaming on platforms like Twitch and YouTube. One such gamer/streamer, a crown-wearing pig avatar known as Technoblade, drew 14.8 million subscribers to his channel on YouTube.

Techno's last *Minecraft*-play video was posted in spring of 2022, drawing 13 million viewers. Just short of three months later, a last video, "so long nerds," was uploaded to his account. The video, posted by his father, explained that Technoblade had died at the age of 23 after a year-long battle with cancer. It drew 74 million views. The video conveyed Techno's feelings about his *Minecraft*/YouTube community journey:

> If I had another hundred lives, I think I would choose to be Technoblade again every single time as those were the happiest years of my life. I hope you guys enjoyed my content and that I made some of you laugh. And I hope you all go on to live long, prosperous and happy lives because I love you guys. Technoblade out.[9]

In an unusual move for a platform personality, Technoblade scrupulously guarded his "real" identity, going so far as to have his brother call him "Dave" on one video as a red herring.[10] His real first name, revealed in his final video, was Alex.

Technoblade started streaming his *Minecraft* play at the age of 13 with a goal of reaching 10 million subscribers. He reached that goal just months before his death. A few weeks later he started hosting "Minecrafters vs Cancer" through YouTube Giving, raising awareness and research dollars for the Sarcoma Foundation of America. He included a donation button on most subsequent videos and livestreams, as well as participating in fund-raising tournaments, ultimately delivering the foundation more than half a million dollars.[11]

Technoblade also used his influencer status to play in a tournament that helped raise more than $340,000 for The Trevor Project, an organization that supports LGBTQ youth.

After his diagnosis, which he announced over a video of *Minecraft* play, Techno encouraged his followers to mask up and get vaccinated to help protect immunocompromised people, like him, from COVID-19.

In short, Technoblade crafted an unusual influencer legacy: Using his platform to do what good he could, for as long as he could.

As with Charlanne Corbin's recognition in *Battleground Europe*, Technoblade was honored by the online community that built up around him.

Following announcement of his death, Techno's fans posted tributes across social media channels, with his handle briefly trending on Twitter. Hypixel, a Minecraft Server Network, donated $50,000 to the Sarcoma Foundation of America and encouraged others to donate in his memory. A digital memory book created by Hypixel drew hundreds of thousands of messages, with this one summarizing the thoughts of many: "Technoblade was important in every aspect of the *Minecraft* community. Rest in peace Techno, you won't be forgotten."[12]

Six days after Technoblade's final video was posted, a new splash text appeared on the 1.19.1 release of *Minecraft*'s Java edition, "Technoblade never dies!"

---

9   From Technoblade's final video, posted to his YouTube channel June 30, 2022.
10 As of this writing, his full name had not been released publicly, but, knowing the internet, it will be (despite the request of his family to honor his wishes and allow his identity to remain confidential).
11 You can check out Technoblade's archived Community posts on YouTube by searching for via his user name. His fund-raising success was noted in an obituary, "YouTube Minecraft player Technoblade—also known as Alex—dies from cancer," aired July 2, 2022, on NPR.
12 Post by nightsky6138 on p. 165 of the "in-memoriam-technoblade" thread on Hypixel.

While his broad fan base may never have known Techno's true identity, the grief was real, particularly for some of his younger fans who discovered him as they worked their way through the unbalancing first two years of the pandemic. One parent, pointing to his sons' difficulty in accepting the loss, noted that Technoblade and others in the online gaming community had become a stable part of their lives in the midst of uncertainty:

> Through this new online, offline world we have, many of them will feel like they will know Technoblade and other members of those online communities better than they know kids in their own school. They will have spent just as much time with them.[13]

One conclusion to be drawn from these two examples, nearly two decades apart, is that relationships built through online gameplay—and the communities that rise up around it—are prevalent, enduring, and very real to those who participate in them. They don't just game; they get to know each other, make friends, tell stories, laugh, get mad, and cry together, all the normal human emotions shared in a community of spirit that transcends time and space but not the fragility of the human body.

Summary

The future doesn't arrive all at once, obliterating the past; innovations emerge and social institutions adjust. The past and future tend to coexist, which means that physical communities and online communities are both seeking our undivided attention. How do we negotiate the tension between our physical selves and online persona? Sometimes we don't; sometimes they exist with remarked separation. Take, for example, Charlanne Corbin, a California woman who built two distinct identities. In her physical life she was a school bus driver known as "Popcorn." Online, she had built a completely separate identity in the massive multiplayer game *World War II: Battleground Europe*. While communities of space (physical) would seem to take precedence over communities of spirit (online), the latter may well be equally real and enduring for the individual.

Key concepts

Antithesis
Avatars
Benign neglect
Common ground
Community characteristics
Exhaustive use
Identification
Intensive use
Lurker
Misidentification
Sympathy
Technopoly
Transcendence

---

13 From "Mourning Technoblade: Fans grieve a Minecraft star they never met" by Amanda Holpuch for *The New York Times*.

References

Block, M. (2022, July 2). YouTube Minecraft player Technoblade—now also known as Alex—dies from cancer. *NPR*. Retrieved from https://www.npr.org/2022/07/02/1109558038/youtube-minecraft-player-technoblade-now-also-known-as-alex-dies-from-cancer

Burke, K. (1968). *Counter-statement*. Berkeley, CA: University of California Press.

Burke, K. (1969). *A rhetoric of motives*. Berkeley, CA: University of California Press.

Holpuch, A. (2022, July 19). Mourning Technoblade: Fans grieve a Minecraft star they never met. *The New York Times*. Retrieved from https://www.nytimes.com/2022/07/19/technology/technoblade-minecraft-tributes.html

CHAPTER 12

Haves, Almost Haves, & Have Nots: The Domestic Digital Divide

Figure 12.1. Digital devices give us direct access to more information than ever before, but the technology underpinning them has significant consequences for those who produce the content that is shared widely across platforms. One result is news deserts—places where local news is no longer covered by journalists— and divisions in public understanding of critical issues. Non Sequitur © 2009 Wiley Ink, Inc. Dist. By Andrews McMeel Syndication. Reprinted with permission. All rights reserved.

Takeaway: *Digital technology access, adoption, and ability are affected by a number of significant and potentially insurmountable issues.*

The motto of the team that built the World Wide Web ultimately became, "if it's not on the web, it doesn't exist."[1] What the programmers were saying was, "upload your ideas into shared space so we can create collaboratively." But in contemporary culture—driven by apps, social media, algorithms, and search engines—this can be more aptly rephrased as, "if you're not online and/or don't have a smartphone, *you* don't exist."

Despite the capacity of the web and social media platforms to bring people together in like-minded communities, there are some who may never connect or spend significant time online for a variety of reasons. This is true in the United States and it's true all over the world. This makes for a digital divide between the technology "haves" and "have-nots," with the latter missing out on a wealth of information and opportunities.

Why does it matter? Those who are online and manage their lives with both computers and mobile devices may have a hard time seeing the problems of those who don't. And if you can't see it, you can't solve it. This is called **marginalization.**

Think about it like shopping for a used print textbook. Someone had it before you and highlighted stuff in three Day-Glo colors, underlined other parts in ink pen, and wrote notes in the page margins. They aren't your notes and, while you might read them, they essentially are a distraction. If you're stuck with a book that someone else has written in, you probably will try to ignore the *marginalia*—the added content that is not part of the main text and that represents someone else's reality.

Our primary cultural text is one steeped in technological utopianism. It's far too easy to shove those who don't share our enthusiasm for technology into the cultural margins and then try to ignore their existence.

The divide between haves and have-nots has existed with the introduction of *all* new technologies. Who had indoor plumbing first? Who had electricity first? Who had televisions and computers in the home and workplace first? The answer is obvious: those who could afford the technology and the range of problems that would come along with its introduction and early use.

Digital technology is different, however, in that a large amount of digital content today is user generated. We are both content producer and content consumer; when everyone contributes, everyone benefits. But there will always be people who don't "get" the nature of the digital world,[2] can't figure out the technology, don't want it, can't afford it, or don't have access to it. And if you can't/won't/don't engage, you're left behind.

While early indications were that technology could solve some of the problems that contribute to the digital divide, in the post-mobile[3] world the divide has grown for a variety of reasons. Because of differences in technology use and adoption, the digital divide must be broken into two parts: the **domestic digital divide** (aka the United States) and the **global digital divide** (aka everywhere else). This chapter is devoted to the former, while Chapter 13 expands on the latter.

---

1   Berners-Lee (2000), p. 163.
2   If you've ever helped your parents or grandparents try to use a smartphone or had to explain social media to them, you understand how frustrating this is for both sides.
3   Mobile applies to smartphones and tablets. Laptops are lumped in with desktops. Think of it this way: If it has a standard physical keyboard, a larger than 15" screen, and multiple connection ports, it's not a mobile device in the same way that a smartphone or tablet is.

The Domestic Digital Divide

Americans tend to believe there is a moral standard for equal entitlement, that everyone should have the same opportunities. It's not true in the United States and it's not true in the world. There are four basic gaps that help explain why the digital divide exists and persists in the United States: technology, information, participation, and influence.

The **technology gap** is due to the uneven distribution of digital communication technologies, which relates to a person's *access* to a given technology. As sci-fi author William Gibson noted in various ways across various interviews in the 1990s, "the future is already here—it's just not evenly distributed."[4]

For example, rural residents of much of the United States don't have reliable cellphone service, much less high-speed broadband internet access. And high-speed, fiber-optic broadband, the fastest internet service possible, is available to fewer than half of U.S. households[5] and offered by few Internet Service Providers.

The District of Columbia, home to the U.S. Capitol, has the second-highest fiber-optic internet access in the nation,[6] and nearly 100% access to very fast broadband internet. That's significant because members of Congress and their support staffs likely encounter nothing but high-speed internet where they use it the most—at work. It also means they're not as likely to understand the need or legislate for digital equity.

The lowest fiber-optic internet coverage is in states with the lowest population, income, and/or education levels, particularly in the rural West. Alaska is at the very bottom of the list, with only about 11% of the population having access to fiber-optic internet.

Google Fiber offers the fastest and least expensive fiber-optic connection, but it's available in fewer than 10 states—sometimes only for residents of a single block—and covers about 1% of the U.S. population. Xfinity (Comcast), Verizon, and AT&T, the largest nationwide broadband suppliers, each offer speeds similar to Google Fiber, but at double to triple the cost.[7] And all have holes in their national coverage, particularly in rural areas.

A key issue of the technology gap is that, even if you are willing to pay Xfinity $300 per month for fiber-optic access, it may not be available where you live. Even if you want it, *you can't have it.*

The **information gap** is due to uneven *adoption* of technology, which relates to a person's desire to use a given technology. It isn't enough to simply have access to information technology, like computers, tablets, and smartphones. You have to think the technology is:

1. *Usable.*
    You know how to use it and can troubleshoot the problems that inevitably arise.
2. *Useful.*
    You see it as useful in *your* everyday life, at work and at play.
3. *Fast enough.*
    The hardware and connection you have are fast enough to provide you with a satisfactory journey into the digital world.
4. *Affordable.*
    Digital devices are expensive.[8] Beyond the device itself, you have to be able to afford the things that go along with it, including connection fees, contracts, taxes, etc.

---

4   Gibson coined the term "cyberspace" in his 1984 novel Neuromancer, which also influenced a range of later sci-fi content, from The Matrix to *Snowcrash.*
5   "Fiber-optic internet in the United States" by Tyler Cooper, BroadbandNow.
6   Nearly 75% of the D.C. has population has access to fiber-optic internet, according to the previous source.
7   "The fastest internet providers in 2022" by Peter Holslin, HighSpeedInternet.com.
8   And they break if you drop them, which makes them really expensive.

There are some well-intended, not-for-profit refurbishing programs that aim to assist with the expense of the device. These typically invite people to donate their old communication technology when they upgrade. The computers and other devices are then cleaned out, spiffed up, and given to families that can't afford newer versions.

It sounds like a good plan, but think of the reasons you upgrade: Your laptop is sluggish from files and viruses, the drivers are too slow for games and video, you've dropped it, and the "E" and "R" keys are missing.

While it's generous to give to others in need, you're giving the device away because it no longer meets your needs. Expert refurbishing helps, but the person who gets your old PC laptop is not going to have the same experience you will with your shiny new MacBook Air.

That's compounded by the fact that about 30% of U.S. households have no broadband internet connection and 11% have *no* access to hard-wired internet at all.[9] Those numbers quickly go up when things like household income and education level are factored in. As school districts nationwide learned at the start of pandemic lockdowns in 2020, a donated laptop with no-to-slow internet and no one on hand for basic tech support is a well-intentioned paperweight.[10]

We're Not Saving the Planet

A significant concern here is that frequent technology upgrades tend to contribute to already colossal piles of e-waste. All electronics produced today contain a range of heavy metals that will contaminate both the ground and water as they degrade.

You have at least three options here:

1. Don't throw the baby out with the bathwater. Try to upgrade the hardware you already have, rather than buying completely new devices.

2. If the hardware is relatively new and you just want something *newer,* take the device to a not-for-profit refurbishing program.

3. If your electronic device really is of no use to anyone, take it to an EPA-certified electronics recycler.

Differences in access to up-to-date devices and high-speed internet make for very different technology experiences. Pretend you're going to Disneyland for Thanksgiving week, to take in the holiday roll-out. Consider the difference between staying at a Motel 6 near Disneyland versus staying at Disney's Grand Californian Hotel & Spa in the park. How might each affect your overall experience?

Of course, staying *in* the park is going to be the most convenient and most fun. But can you afford it? The Motel 6 closest to Disneyland is about $139 per weeknight (for a family of four, not including lodging taxes). One eight miles away is about $85 per night. The least expensive room at the on-site Grand Californian is a relatively staggering $924 per weeknight, not including park tickets. Choosing the latter would be impossible for some families.

To bring this back around to digital technology, the haves can afford the best experience, from high-speed broadband to fiber optics. The have-nots, relegated to dial-up or bargain-speed broadband, are left to wonder what everyone is so excited about.

9   "Internet subscriptions in household," American Community Survey, U.S. Census.
10   "Pandemic response lays bare America's digital divide" by Victoria Bekiempis, The Guardian.

Other barriers to adoption relate to the nature of the technology itself. If you believe a technology exists solely for entertainment (such as gaming consoles and smart speakers), then the information gap can be particularly wide. The barrier here is largely desire—"I don't *want* that."

The information gap narrows significantly when a technology reaches the point at which it fulfills a critical need, such as access to governmental, occupational, educational, or medical services.[11] The home telephone eventually emerged as such a technology, but the web has quickly evolved into something people *need*—depending on what it is being used for. Job applications? Check. Income tax returns? Check. College applications and community college enrollment? Check.

As smartphones have reached near saturation level of adoption in the United States, it's become an expectation for people to have them to communicate for both work and play. Virtually all workers at every level of the economic food chain are expected to be accessible, and that accessibility often translates into smartphones and communication connectivity. Lower-income service workers, who aren't as likely to have home internet or an email address they regularly check, may in fact have greater demand for cellphone access from employers and potential employers. Technology researcher Julia Ticona argues that this requirement effectively has become a new tax on the nation's poor:

> Amid historic levels of income inequality, phones and data plans have become an increasingly costly burden on those who have the least to spare. … The cost of connectivity represents more than half of what these households spent on electricity, and nearly 80 percent of what they paid for gas. As a proportion of household income, the lowest earners spent four times more on phones than high earners.[12]

Ticona noted that customers who can't afford the up-front cost of a phone and service are often forced to pay more predatory via third-party financing. And simply being connected is just a lot more effort when you don't have money to spare:

> From juggling broken phones, borrowing money to pay bills, and, when all else fails, finding free sources of internet, the tenuous grasp these workers have on connectivity is maintained through constant effort. Many rely on coffee shops and fast-food joints in their neighborhoods and throughout their commutes, facing racist threats and harassment from managers and even the police while trying to swap a shift through an app, download music for a long shift, or text their boss that they were running late.[13]

Ticona's concerns are reflected in efforts by some lawmakers to frame connectivity as a luxury, rather than the necessity it has become. For example, U.S. Rep. Jason Chaffetz (R-Utah) argued in 2017 that if Americans can afford iPhones, they should instead spend their money on healthcare. As he noted in interview with CNN: "Americans have choices, and they've got to make a choice. So rather than getting that new iPhone that they just love and want to go spend hundreds of dollars on that, maybe they should invest in their own health care."[14]

The **participation gap** comes when people have access to technology and adopt it to some extent but don't achieve the *ability* to participate in the digital conversation.

---

11   Samoriski (2002).
12   "Smartphones are a new tax on the poor" by Julia Ticona, Wired.
13   Ibid.
14   "The luxury of telling poor people that iPhones are a luxury" by Brian Fung, The Washington Post.

Web usability guru Jakob Nielsen[15] referred to this as the "empowerment divide" and predicted that rapid and continual technological innovation has the capacity to make this much worse in the future:

> The internet can be an empowering tool that lets people find good deals, manage vendors, and control their finances and investments. But it can just as easily be an alienating environment where people are cheated. Members of the internet elite don't realize the extent to which less-skilled users are left out of many of the advancements they cheer and enjoy.[16]

The basic concern is that even if contributing to the digital conversation is as easy as logging into a social media site and making a post, many people still would find no other use for digital media than looking at other people's content.

Nielsen pointed to a 90/9/1 equation: 90% of web users don't contribute original digital content; 9% contribute as it suits them; and 1% dominates content production. While Nielsen made this observation in 2006, repeated studies show the problem persists.

Nielsen attributed this, in part, to large-scale **digital *illiteracy***—not knowing the difference between a browser and a search engine or being able to distinguish news from gossip.

There are plenty of smartphone users who don't understand how the internet and web are structured, how search engines find content, what user agreements mean, the differences among apps, and/or how ads and feeds are personalized. And this is not an issue that relates to age. It's not just the "oldens" who are digitally challenged; **digital natives**[17] may be a little worse because they didn't witness the evolution of digital technology and don't question what they're consuming or how it consumes them.

What do most people actually do online? They check email, surf the web, fact-check a wide range of random stuff,[18] skim headlines, scroll through social media feeds, bank, pay bills, book travel, game, and shop. What most people are not doing is uploading original content outside of social media—videos, comments, wiki contributions,[19] and so on.

Not knowing how to navigate digital offerings makes people more vulnerable to misinformation, disinformation, scams, phishing, and identity theft. Because people often know enough to recognize the risk, they are wary of using web resources that can be genuinely beneficial for sharing original content that might connect them to others—the optimal use of social media.

Sometimes, it's a lack of initiative, not skill, that is the problem: It's just easier to use the web browser that came with a computer or smartphone than to spend five minutes downloading a different browser with stronger privacy settings. As Nielsen warned, "this means that the user's attention can be sold off like a sheep to slaughter"[20] by the companies that own the websites and apps they frequent.

---

15   "Digital divide: The 3 stages" by Jakob Nielsen, Nielsen Norman Group.
16   By the way, anything involving Jakob Nielsen and communication technologies—whether online or in print—is worth consuming. He's head of the Nielsen Norman Group, a usability consulting firm with headquarters in California. His "Alertbox" articles are the best. Really.
17   Those born between 1995 and 2010, who have not experienced life without the World Wide Web.
18   Who's the tallest president? What year did the song "YMCA" come out? What does a brown recluse spider look like? You know you want to know these things (except that last one—that might be an urgent need).
19   A wiki is a crowdsourced site, which means that user contributions make up the whole. The best example is Wikipedia. Wiki also is a Hawaiian phrase for hurry up, get it moving now. Wiki-wiki! The rest of the chapter awaits!
20   Nielsen, op. cit.

Google Chrome, for example, has been classified as spy software by many tech journalists. Firefox, on the flip side, is owned by the nonprofit Mozilla Foundation, which has a greater dedication to digital privacy and less incentive to market your personal data to advertisers. Consumer tech journalist Geoffrey Fowler discovered the following in a comparison study of the two web browsers:

> My tests of Chrome vs. Firefox unearthed a personal data caper of absurd proportions. In a week of Web surfing on my desktop, I discovered 11,189 requests for tracker "cookies" that Chrome would have ushered right onto my computer but were automatically blocked by Firefox. … Chrome is even sneakier on your phone. If you use Android, Chrome sends Google your location every time you conduct a search. (If you turn off location sharing it still sends your coordinates out, just with less accuracy.) [21]

And Gmail, also owned by Alphabet, automatically logs you into Chrome, which expands access to your data. Social media sites are no better, with virtually all profit-driven sites selling user demographics and/or data to advertisers.

The **influence gap** is a result of more than a decade of ubiquitous social media, which allows people to avoid content that might otherwise challenge their perspectives. That ability not only brings together those of like minds, but also sows seeds for ***division***. The result is **algorithmic discord**, which allows people effectively to create and dwell in alternate, often oppositional, realities.

Not only do we choose what content we want to see and who can see ours, but social media algorithms helpfully find posts that relate to our interests. This not only protects us from alternate views, but efficiently reinforces existing values and beliefs. There's a word for this: homophily, which neatly translates into a proverb used to explain human behavior since the 1500s: "birds of a feather flock together."

Social media companies didn't start out with a mission for division. Instead, the grand design may indeed have been to help us connect with each other. The baseline desire for profit ultimately tilted the companies toward microtargeting and strategies to keep us on social media sites longer. That meant algorithms evolving in ways that hone in on our interests and connected us to others who share them—most importantly advertisers with related products who pay social media companies for access to our exclusive digital space.

By 2012, when smartphones were taking hold and delivering Facebook to our fingertips, technology researchers began to note a major shift in the influence of social media. Algorithm-driven recommendations just made things worse. One result was that misinformation could run rampant as the technology helped sort people into like-minded tribes. The loss of common ground also increased the potential for abuse and manipulation. As sociologist Zeynep Tufekci noted:

> It was a shift from a public, collective politics to a more private, scattered one, with political actors collecting more and more personal data to figure out how to push just the right buttons, person by person and out of sight.[22]

Tufekci makes the important point that the problem is not that we're only exposed to views like our own online. In truth, we have access to a larger range of information and opinions than at any other time in history. But the way we engage with information that contradicts our own beliefs has shifted from consideration to competition. Tufekci said it's a classic case of in-group versus out-group, us versus them: "It's like hearing them from the opposing team while sitting with our fellow fans in a football stadium. … Belonging is stronger than facts."[23]

---

21   "Goodbye, Chrome: Google's web browser has become spy software" by Geoffrey A. Fowler, *The Washington Post*.
22   "How social media took us from Tahrir Square to Donald Trump" by Zeynep Tufekci, *MIT Technology Review*.
23   Ibid.

Pandemic isolation accelerated the decline of shared physical space, resulting in a corresponding rise in selective virtual space and the potential for algorithmic discord.

Social media constrains us to the rules of the algorithm. We don't have to engage with or even acknowledge things that don't fit our worldview. Instead of listening to and considering other perspectives, we can mute, block, unfollow, and report—effectively canceling those who post them. Worldviews are narrowing and divisions broadening, with virtually all of our traditional social institutions are bending from the effects.

It's a technology-imposed evolution that Marshall McLuhan predicted nearly 60 years ago:

> The medium, or process, of our time–electric technology–is reshaping and restructuring patterns of social interdependence and every aspect of our personal life. It is forcing us to reconsider and reevaluate practically every thought, every action, and every institution formerly taken for granted. Everything is changing–you, your family, your neighborhood, your education, your job, your government, your relation to "the others." And they're changing dramatically.[24]

The Influence of the Demographic Divide

Early concerns were that gender and race would be factors in determining technology adoption and participation. That's because middle-aged white men developed the World Wide Web and dominated its usage early on. More than 30 years after its invention and expansion of digital influence via smartphones, most studies show that it's connectivity that's the issue. Once people are engaged with the web and mobile apps, they tend to use them—a lot—regardless of race or gender. But there are other factors that come into play in this complex equation.

Woven throughout the technology, information, participation, and influence gaps are four demographic characteristics that work together and apart to reduce digital technology engagement: Age, income, education, and location.

About 7% of U.S. adults ages 18 and up don't go online at all, according to the Pew Research Center.[25] While 7% doesn't sound like much, it's more than 23 million people who are not engaged in the digital conversation. These statistics and the discussion that follows may help explain why:

Age.   Nearly 100% of those under 29—those born after invention of the World Wide Web—report at least some internet use. Perhaps equally unsurprising, the lowest use is in the 65 and older age group, with 75% reporting internet use.[26] Only 20% of that age group has a computer of any sort in the home.[27]

Income.   Some 99% of those with an income of $75,000 and above routinely go online, as compared with 86% for those with an income of less than $30,000.[28] Overall, only about 85% of American households subscribe to a home internet service. For households with income of $20,000 or less, that connectivity rate plummets to 36%. For households with income of greater than $75,000, 96% have home internet service.[29]

Education. Of American adults with a bachelor's degree or higher, more than 93% have at least one computer and smartphone in the home and 94% have home broadband service. Of those who don't complete high school, only 45% have a computer, 67% have a smartphone, and 63% have home broadband service.[30]

24   McLuhan and Fiore (1967), p. 8.
25   "7% of Americans don't use the internet. Who are they?" by Andrew Perrin & Sara Atske, Pew Research Center.
26   Ibid.
27   "Computer and Internet use in the United States" by Michael Martin, U.S. Census Bureau.
28   Perrin & Atske, op. cit.
29   "Types of computers and internet subscriptions," American Community Survey, U.S. Census Bureau.
30   Martin (2021), op. cit.

Location.   By region, digital connectivity in the U.S. is highest in urban and suburban areas (94%) and lowest in rural areas (90%). Connectivity overall is highest in the West (94%) and lowest in the South (91%).[31] Ten states have broadband internet—the most common form of connectivity in the U.S.—available to less than 80% of the population.[32] Two states, Mississippi and Montana, are ranked in the bottom 10 for internet connectivity, speed, and cost, while Massachusetts and Rhode Island are ranked in the top 10 for all three.[33]

What's our takeaway from all this?

1. **The United States has achieved internet saturation.**

   There will always be people who do not use digital communication tools, whether by choice or circumstance. That has significant implications for investment in the U.S. information infrastructure, which is funded by corporations.

   Remember our discussion of the internet and World Wide Web in Chapters 9 and 10? Private enterprise maintains the submarine cable system, the underground lines connecting the internet from coast to coast, and the extremely expensive data centers that keep content flowing. If 93% of Americans are already online, but less than 70% of the population of China and less than 55% of India are online,[34] where would you expect publicly held companies[35] to put their money? The somewhat obvious answer: Elsewhere, capitalizing on growing international markets in developing countries.

   It's a matter of math: 7% of the American population is about 23 million people; tapping into the remaining 30% of the Chinese population not online is a market of about 448 million people. For India, that remaining 45% is about 640 million people.

2. **Retirement happens.**

   Once people reach retirement age, there's a steady drop-off in digital engagement—with good reason. If you have a job that requires work online—including reports, email, texting, and video teleconferencing—you can expect that day in and day out for your entire career.

   If you start your first post-college job at 23 and retire at 65, that's 42 years of email and dozens of upgrades, system conversions, innovations, and inventions: in short, *endless change*. Digital engagement use is highest in peak earning years and lowest after the age of retirement. This drop-off is expected to continue, as an inevitable degree of technology fatigue sets in after decades of work.

3. **Demographics influence each other.**

   The digital divide is underpinned by interlinking demographic factors that create a hard-to-break cycle—one that more technology can't fix.

   The Showtime original series "Shameless" had a spot-on depiction of this reality in its first season.[36] The show chronicles a family of eight, living below the poverty line in South Side Chicago, an area known for its high crime rate. Both parents are largely absent—struggling

---

31 Perrin & Atske, op. cit.
32 The states with the lowest broadband connectivity: Alabama, Alaska, Arkansas, Hawaii, Louisiana, Mississippi, Montana, Oklahoma, Vermont, West Virginia.
33 "Best and worst states for internet coverage, prices and speeds" by Tyler Cooper & Julia Tanberk, BroadbandNow.
34 For current numbers, go to internetworldstats.com.
35 Think Comcast, Verizon, AT&T, Alphabet, Amazon, Apple, Meta, Microsoft, and so on. Most already are seeking investment in developing countries, not overdeveloped ones.
36 "It's time to kill the turtle," season 1, episode 8.

with mental health and addiction—while eldest sister Fiona drops out of high school to try to keep the family together.

Fiona works temporary evening and weekend jobs, but longs for the stability of a steady office job. She attends a free class to learn basic office skills, and the instructor leaps right into creation of a PowerPoint presentation. Fiona's dated laptop slowly boots up and she struggles to find and open the program. She quickly realizes she doesn't have the minimum understanding of professional technology use that's shared by everyone else in the room. Her confusion morphs quickly from frustration to anger and she gets up and leaves. Hours later, she's back at her Hooters-style bar job.

Fiona's digital engagement was shaped by her household income, education, and location, which effectively kept her from gaining skills needed to succeed outside of the service industry.

4.   **Internet infrastructure is expensive. So's connectivity.**
     Residents of lower income areas are less likely to connect to the internet. Internet Service Providers tend to wire relatively wealthier areas first. They're looking for customers and have less incentive to go into poorer areas. That leaves out inner cities.

     Providers also tend to bypass many of the rural areas in the United States. That's because it's more expensive to connect across distance for fewer people. Inner cities and rural areas often also have lower incomes and lower education rates, which means that people may not connect, even if access is available.

     This is where America's reliance on a profit-centered technology industry to provide connectivity becomes a problem. The solution? Broadband equity—fast internet for all—as an expectation of the government.

5.   **Digital technology use is a choice.**
     With ever-accumulating technology costs, people often must choose what they can afford. For example, 11% of Americans over the age of 18 access the internet through their smartphones, with no hardwired service at home.[37] Some 66% of those who are not currently online, but have been online in the past, said it is unlikely they will return to internet usage in the next year—particularly if information technologies have little relevance to their work and personal lives.[38]

6.   **Mobile devices are widening the domestic digital divide.**
     Being able to get online is not the problem. Personal computers are relatively inexpensive and there are public spaces, like libraries, where computer use is free.[39]

     But what about not having a smartphone or tablet at your fingertips in our increasingly mobile-centric culture? Think of it this way: If you can't afford to pay for the latest, greatest smartphone and the monthly fees that go along with it, you're a have-not.

As digital media have evolved, critics have claimed that the gap between the haves and the have-nots is widening in the United States. Media scholar Jan Samoriski said this about the problem:

---

37   "Types of computers and internet subscriptions," op. cit.
38   Cord cutting = freedom. From the 16th annual study by the Digital Future Project on the impact of digital technology on Americans, "Surveying the digital future."
39   Except during pandemic lockdowns, which also closed libraries and left those dependent on them for digital connectivity without that option.

Not being online sets in motion a series of events similar to the cycle of poverty. Without access to computers and the Internet, the digitally disadvantaged are denied the opportunity to acquire the knowledge and skills they will need to get jobs in the information industry. They are therefore unable to compete for the higher-paying technology jobs and enter the occupations that would allow them to escape the cycle. Communities also suffer because they cannot provide employees to fill high-tech jobs, which is a prerequisite for attracting high-tech industries.[40]

Other Factors That Influence Digital Engagement

There's a minor rebellion going on among people who can afford the technology and are well-educated but simply choose not to engage with it. These people are **neo-Luddites**.

Luddites, from which the term has evolved, were people—primarily artisans—who fought the industrial revolution in Great Britain. The Luddite movement began in the early 1800s and so-called Luddites destroyed industrial-sized looms that were created to make textiles. In other words, they were threatened by the technology and the effect they perceived it having on society.

"Neo" means "new," so a neo-Luddite is a person who thinks modern technologies are corrupting society, particularly information communication technologies like television, the Internet, and mobile.

Most neo-Luddites are people who simply choose not to adopt new technologies and, when they are forced to use things like computers for work, they use them for work; they don't welcome them into their personal lives. They are fighting against what may best be termed the ubiquitous tyranny of technology.

There are also people who become de facto neo-Luddites, which means that they reach a point in their lives/careers where technology fatigue has begun to set in.[41] They simply stop adopting and adapting, preferring to let others lead the way.

The opposite of a neo-Luddite is an **early adopter.** These people continually watch for new technologies, research them, and buy them as soon as they are on the market (e.g., people who stand in line at the Apple Store for 24 hours to get their hands on the latest, greatest iPhone). Early adopters tend to have a more **utopian** perspective on technology—believing that digital engagement is making the world better.

Going back to our earlier discussion of demographics, who do you think might be early adopters? Overly simplified answer: younger people, in cities, who are well-educated and have disposable income.

Somewhere in between these two are **technorealists,** people who understand that technological change creates undesirable as well as desirable possibilities.

In short, even those who have the desire, access, and ability to use digital information technologies don't engage with them to the same degree. Although roughly 93% of the U.S. population is online, many aren't using that access consistently, collaboratively, or beyond work requirements.

We'll now turn our attention to the second part of the digital divide, global engagement.

Summary

According to the founding ideology of the World Wide Web, information housed on the hard drives of individual computers are useless in a world of collaborative knowledge. What we don't *share,* others can't *know.* Individuals have little impact in digital isolation but can be profoundly influential when engaged online. But there are a variety of reasons why people are not online and may never contribute original content to the knowledge base facilitated through broad digital engagement. This is due to

---

40   Samoriski (2002), op. cit.
41   Just like wearing a muumuu pegs you as somebody's nana, having a Bluetooth phone earpiece in your ear pegs you as post-60 and at least two decades beyond cool.

gaps in technology (access), information (adoption), participation (ability), and influence (algorithmic discord). Factors such as age, income, education, and location can greatly impact how, or even if, people engage with digital technologies in the United States.

Key concepts

Algorithmic discord
Demographic divide
Digital illiteracy
Digital natives
Domestic digital divide
Early adopters
Influence Gap
Information Gap
Marginalization
Neo-Luddites
Participation Gap
Technology Gap
Technorealists
Utopian

References

Bekiempis, V. (2020, March 21). Pandemic response lays bare America's digital divide. *The Guardian*. Retrieved from: https://www.theguardian.com/world/2020/mar/21/coronavirus-us-digital-divide-online-resources

Berners-Lee, T. (2000). *Weaving the web: The original design and ultimate destiny of the world wide web by its inventor*. New York, NY: HarperCollins.

Cooper, T. (2022, April 12). Fiber-optic internet in the United States. *BroadbandNow*. Retrieved from https://broadbandnow.com/Fiber

Cooper, T. & Tanberk, J. (2022, May 6). Best and worst states for internet coverage, prices and speeds. *BroadbandNow*. Retrieved from https://broadbandnow.com/research/best-states-with-internet-coverage-and-speed

Fowler, G. A. (2019, June 21). Goodbye, Chrome: Google's web browser has become spy software. *The Washington Post*. Retrieved from https://www.washingtonpost.com/technology/2019/06/21/google-chrome-has-become-surveillance-software-its-time-switch/

Fung, B. (2017, March 8). The luxury of telling poor people that iPhones are a luxury. *The Washington Post*. Retrieved from: https://www.washingtonpost.com/news/the-switch/wp/2017/03/08/the-luxury-of-telling-poor-people-that-iphones-are-a-luxury/

Holslin, P. (2022, July 1). The fastest internet providers in 2022. *HighSpeedInternet.com*. Retrieved from https://www.highspeedinternet.com/resources/fastest-internet-providers

Internet subscriptions in household. (2019). *American Community Survey*. Retrieved from https://data.census.gov/cedsci/table?q=ACSDT1Y2019.B28011&tid=ACSDT1Y2019.B28011&hidePreview=true

Martin, M. (2021, April 21). *Computer and internet use in the United States, 2018. U.S. Census*. Retrieved from https://www.census.gov/content/dam/Census/library/publications/2021/acs/acs-49.pdf

McLuhan, M. & Fiore, Q. (1967). *The medium is the massage: An inventory of effects*. New York, NY: Bantam Books.

Nielsen, J. (2006, November 20). Digital divide: The 3 stages. *Nielsen Norman Group*. Retrieved from http://www.nngroup.com/articles/digital-divide-the-three-stages/

Perrin, A. & Atske, S. (2021, April 2). 7% of Americans don't use the internet. Who are they? *Pew Research Center*. Retrieved from: https://www.pewresearch.org/fact-tank/2021/04/02/7-of-americans-dont-use-the-internet-who-are-they/

Samoriski, J. (2002). *Issues in cyberspace: Communication, technology, law, and society on the Internet frontier*. Boston, MA: Allyn & Bacon.

Ticona, J. (2021, December 16). Smartphones are a new tax on the poor. *Wired*. Retrieved from: https://www.wired.com/story/phones-connectivity-tax-policy/

Tufekci, Z. (2018, August 14). How social media took us from Tahrir Square to Donald Trump. *MIT Technology Review*. Retrieved from https://www.technologyreview.com/2018/08/14/240325/how-social-media-took-us-from-tahrir-square-to-donald-trump/

Types of computers and internet subscriptions, American Community Survey, U.S. Census Bureau. https://data.census.gov/cedsci/table?q=broadband&tid=ACSST5Y2020.S2801

CHAPTER 13

Connectivity & Digital Disruption:
The Global Digital Divide

Figure 13.1. Our unquenchable thirst for connectivity has consequences. Over the Hedge © 2020 M Fry & T Lewis. Reprinted by permission of Andrews McMeel Syndication for UFS. All rights reserved.

Takeaway: *The country-by-country approach to connectivity and information makes the global digital divide all but impossible to bridge, with the same lack of consensus and cooperation increasing threats to cybersecurity.*

On a cloudless day, look up at the sky. What do you see? Odds are there are puffy white, crisscrossing remnants of jet contrails.

Now, imagine it's the 1950s and public jet travel is rare, but gaining ground, in the United States. Perhaps you've never been on a jet and are not entirely sure about the idea of that mode of passenger transport, then you look up and see … crosses in the sky.

There is anecdotal evidence of people believing that crossing contrails signaled end times and urgently contacting public officials and news reporters.[1] Seems like a quaint story today, when it's hard to imagine life without air travel or a contrail-free sky. In fact, that was one of the most unnerving things about both September 11, 2001, and the early pandemic lockdown in the U.S.—no air travel, no contrails.

Now, find a dark spot some late evening and stare at the night sky. If you're patient, you might be rewarded with something that today may feel as jarring as the early contrail crosses: A *moving line of stars*, evenly spaced with an odd hole here and there.[2] One tech journalist described the line as "a chain of fairy lights zooming across the night sky."[3] It begins and, eventually ends. Of course, they're not stars, which move position with the Earth's rotation, they're satellites.

The Quest to Connect the Planet

While there are thousands of satellites in the Earth's orbit, they tend to be deployed individually for particular purpose, not connected in a ribbon speeding across the night sky. That string of aligned satellites is part of Starlink, a mobile, high-speed, low-latency internet service aimed at closing the internet divide in rural and isolated areas, like ships at sea.[4] Starlink was launched in 2020 by SpaceX, a space exploration company owned by Elon Musk.

As of mid-2022, SpaceX had deployed more than 2,500 Starlink satellites into the Earth's orbit, with plans for thousands more.[5] To put that in perspective, prior to Starlink launches, there were about half that many functioning satellites orbiting the planet, with just over 12,000 satellites launched since global space programs began.[6]

SpaceX alone has U.S. Federal Communications Commission approval for 12,000 satellites, with another 30,000 requested internationally, to create a "megaconstellation" of 42,000 satellites.[7]

Of those launched by SpaceX to date, several hundred have failed or fallen from orbit, causing the gaps in the link. For example, of 49 Starlink satellites sent into orbit February 3, 2022, 40 were knocked out by a geomagnetic storm—caused by a solar storm—within 24 hours.[8] A video posted on

---

1   True story. Author Wiesinger's mom was a reporter at the *San Angelo Standard-Times* in 1955 and vividly recalls a terrified man running into the newsroom, shouting about crosses in the sky.

2   If you want to see what the chain of satellites looks like for yourself, go to findstarlink.com and put in your location. You also can find videos on YouTube.

3   "A SpaceX fanatic created a website to find out when Starlink satellites were visible in his location" by Kate Duffy, *Insider*.

4   Fun fact: Starlink Maritime launched in July 2022 at a cost of $5,000 per month, plus an initial $10,000 fee for two satellite dishes. It's for yachts. (Of course it is.)

5   "SpaceX passes 2,500 satellites launched for Starlink internet network" by Stephen Clark, Spaceflight Now.

6   "SpaceX Starlink internet: Costs, collision risks and how it works" by Adam Mann, Tereza Pultarova, & Elizabeth Howell, Space.com.

7   Ibid.

8   "Geomagnetic storm and recently deployed Starlink satellites" from SpaceX Updates.

YouTube shows fireballs in the nightsky over Puerto Rico, as the doomed satellites were incinerated in the Earth's atmosphere.[9]

As of this writing, Starlink has slower upload and download speeds than hard-wired broadband internet, but is providing reliable internet to areas unaccustomed to it. The service has a competitive month-to-month fee, but requires an up-front investment in a Starlink satellite dish, a user terminal nicknamed "Dishy McFlatface," which has a long customer wait list to purchase.[10]

Starlink is one of a handful of projects by various tech companies and entrepreneurs trying to expand opportunities for internet service beyond broadband and fiber optical cables, which must have a physical connection to a home or business. At present, those services are offered in the U.S. by a very few companies, notably dominated by Xfinity (Comcast) and AT&T, with slow to no service to many rural areas.

And the companies also are looking beyond the United States, to a largely untapped internet market in rural areas of developing countries, and toward quickly restoring internet service following natural and human-caused disasters.

Starlink, for example, was activated in Ukraine two days after the February 24, 2022, invasion of Russian troops—a development Elon Musk announced with a tweet. Within three months of the war's start, there were more than 10,000 "Dishy McFlatface" terminals in place, taking the place of a largely destroyed internet infrastructure. The impact spanned from public safety to military use.[11]

In a plan largely mirroring Starlink, Amazon announced in late 2021 it had partnered with Verizon on Project Kuiper, a high-speed broadband and 5G wireless communication system aimed at boosting global internet speeds and connectivity. The venture will deploy and network more than 3,000 satellites in low earth orbit.

Rockets owned by Blue Origin, an aerospace company founded by Amazon's Jeff Bezos, were to be among those expected to begin deployment of Project Kuiper satellites in late 2022.

Starlink and Amazon's Project Kuiper also shared in hundreds of millions of dollars in NASA funding aimed at creating new opportunities for space communication. While both companyies intend to use the satellites for Earth-bound communications, as well, the U.S. government is hoping to replace the aging satellite systems it owns and operates.[12]

Both Starlink and Project Kuiper have drawn criticism for a number of reasons, including concerns about accumulating space junk and light pollution.

The proliferation of satellites increases the possibility of them crashing into each other and other space debris ringing the planet. And more space debris begets more space debris, accelerating the pace of trash accumulation in low-Earth orbit. That cascade of collisions, known as **Kessler Syndrome**, would lead to enough space debris to disable all satellites and effectively block our ability to safely leave orbit.[13]

A second concern is that Starlink satellites are powered by solar panels that reflect the sun and, despite efforts to darken the shiny parts, a megaconstellation of them would have the capacity to completely reshape our vision of the night sky. Not only would they obscure and outshine stars visible to the naked eye, but they would create blind spots for astronomers watching the sky for approaching meteors.

---

9   Check the video out here https://youtu.be/mUlAz_Oxv4Q
10   "Elon Musk, are you out there? SpaceX Starlink internet service gets more reliable," by Todd Bishop, *GeekWire*.
11   "How Elon Musk's Starlink go battle-tested in Ukraine" by Vivek Wadhwa & Alex Salkever, Foreign Policy.
12   "NASA awards millions to SpaceX and Project Kuiper for satellite communications" by Alan Boyle, GeekWire.
13   This process of "collision cascading" was predicted in 1978 by NASA scientist Donald Kessler, who thought it would take 30 or 40 years for the trash to accumulate to the point where it made space travel impossible. Satellite numbers, however, remained fairly low—until now.

Astrophysicist Paul Sutter described the potential impact of Starlink satellites in the night sky like this:

> When a satellite constellation crosses a telescope's field of view, it isn't just a single streak but multiple ones that can potentially wreak havoc on astronomical observations. … It's impossible to know just how bad the skies will get until all the satellites are up and astronomers try to do astronomy. But by then, it might be too late.[14]

Other Approaches to Global Connectivity

Previous Big Tech projects aimed at closing the global digital divide include Alphabet's Project Loon, which proposed a grid of silvery balloons hanging in the stratosphere to provide global internet connectivity. It was launched in 2011 by Alphabet's Moonshot Factory.

Project Loon was deployed to Puerto Rico in November of 2017 to restore internet and mobile phone service following extensive damage to the island from Hurricane Maria. While such an effort would normally take months to years, Project Loon restored connectivity within a couple of weeks.[15]

Loon also provided internet to areas of New Zealand, Kenya, and other underserved areas before being scuttled in 2021 when it was determined to have no commercial viability.[16]

Not long after Project Loon began, Mark Zuckerberg put out a plan for Facebook to provide universal basic internet to the world. In 2013, he launched Internet.org, which proposed a network of satellites, solar-powered planes, drones, and lasers to provide internet and phone connectivity.[17] While the Facebook CEO intended the project as a humanitarian effort, it was met with skepticism from the start and, eventually, outright protest:

> At one point, 67 human rights groups signed an open letter to Zuckerberg that accused Facebook of "building a walled garden in which the world's poorest people will only be able to access a limited set of insecure websites and services."… While the app allowed numerous services, they were concerned that Facebook was the ultimate arbiter of which ones were included. Facebook had much to gain by centralizing the web onto one platform: Facebook.[18]

Initial deployment of Internet.org to India was put on hold—and ultimately rejected—when it turned out the project would only provide free access to Facebook products and partner services. After increased pushback by various world governments and human rights organizations, Zuckerberg renamed Internet.org to Free Basics in an effort to help people understand the utility of service, opened up the application to more services, and improved security.[19]

Unlike Project Loon, Free Basics—now part of Meta Connectivity—has enjoyed moderate commercial success, connecting an estimated 100 million people across dozens of countries to internet and phone service via its Discover app.

Microsoft launched its own internet project in 2017, the Microsoft Airband Initiative, aimed at reducing the digital equity gap by bringing broadband connectivity to millions of rural Americans.[20] Framed as part of Microsoft's corporate responsibility, the five-year project forged partnerships with existing internet and phone service providers, hardware and software companies, and state and local

---

14   "Megaconstellations could destroy astronomy and there's no easy fix" by Paul Sutter, Space.com.
15   "Project Loon team gave Puerto Rico connectivity" by Nathan Mattise, *Ars Technica.*
16   Check out pictures of the Loon balloons and learn more about the project at x.company/projects/loon/.
17   Fun fact: In 2016, Facebook contracted with SpaceX to put the first Internet.org/Free Basics satellite into orbit. It blew up over Africa, for which it had been intended to provide service.
18   "What happened to Facebook's grand plan to wire the world?" by Jessi Hempel, *Wired.*
19   Ibid.
20   Find out more about the Microsoft Airband Initiative at microsoft.com/en-us/corporate-responsibility/airband-initiative.

governments. This collaborative approach elicited little controversy and drew attention to internet and phone service gap areas nationwide.

With Big Tech companies forging the path to greater global connectivity, the ultimate solution likely will require some combination of innovation, collaboration, and sustained attention to corporate responsibility, as well as a more coordinated approach to how new technologies are evaluated for potential harm and regulated internationally.

Governments Have a Key Role in How Technology Spreads

When considering the **global digital divide**, the uneven distribution of communication technology across the planet, the influence of governments within individual countries is paramount. Building from a range of diverse ideologies, governments ultimately make choices about the digital infrastructure, the degree to which information will be controlled, and how the technology will be used for monitoring individual behaviors.

In the United States, for example, innovation to and maintenance of systems like broadband internet connectivity and the 5G wireless communication network are almost entirely relegated to private enterprise—profit-driven, publicly held companies.

The U.S. government has made very little effort to constrain the information that flows across those networks, depending instead on the companies that host the content to police it. This has led to a hodgepodge of approaches, where social media users are put in Facebook "jail" for seemingly minor infractions, while influencers may be monetarily rewarded for the same types of content.

These efforts reflect the two paths countries have in approaching digital engagement for their citizens: infrastructure development and information control.

Infrastructure development is necessary to foster national development. If a country wants to create the perception that it is ready for international investment, it needs to ensure that the physical backbone is in place and people understand how to use it. That means both investment in the technologies that link the country together, as well as development of a culture that embraces digital engagement and technological innovation.

There are a number of reasons why countries don't invest in infrastructure development, many of which relate to poverty, disease, strife, war, vast expanses of sparsely populated land, and so on. For example, government attention to information infrastructure in the United States—a campaign platform of Barack Obama when running for his first term of office in 2008—was disrupted by war on two fronts, political divisions, and global economic meltdown.

While a motivated government can ensure that infrastructure changes take place within a few years, a cultural change can take decades—and both must exist for a country's efforts at technological development to succeed and be worthy of further investment.

Countries may assume full control over the internet infrastructure within their borders to enhance national security, particularly from cyberattacks, or they may opt for temporary control to shut down the internet in the event of civil unrest.

Cuba, for example, has expanded internet availability in the past few years, but cut online access to the entire island for 72 hours in 2021 in response to growing citizen protests. Not just one website or service, the internet.

Information control is how a government approaches digital information—from news organizations to social media sites—as a way to control its citizens. Governments have two general goals when seeking information control:

- maintaining political stability and power by controlling *all* messages, and

- preserving religious, moral, and social values.

Many authoritarian governments choose to fully control infrastructure in order to fully control information access.

North Korea largely limits use of global internet to those charged with national security, prohibiting citizen access entirely. Instead, North Koreans are connected via an intranet, an in-country network with zero out-country access.

Other countries are less concerned about control of the internet infrastructure within their borders and allow foreign investment to build up the systems. To block and censor, they only need leverage over internet and platform service providers—such as the ability of foreign companies to do business in the country at all.

Indonesia allows foreign content, like social media, but requires *all* services to register with the government, turn over customer data on request, and comply with content moderation. Indonesia's regulation of the internet, which is the most restrictive in the Asian Pacific region, was drafted by the government to ensure that digital information is used in a "positive and productive way."[21]

A number of countries view both Western culture and related web and app content as corruptive, and block it accordingly. For example, American legal perspectives on politics, capitalism, education, health care, religion, gun control, drug and alcohol use, women's rights, and LGBTQ acceptance aren't shared globally.[22]

Saudi Arabia and other Arab countries monitor all forms of Western content and restrict much of it on religious and moral grounds, while still facilitating development of infrastructure for high-speed internet and 5G communication.

There also are a relatively large number of countries that don't want Western political perspectives to influence the minds of their people. Those things generally result in a government choosing to limit or block internet content that comes into or leaves a country.

The Beginnings of a Global Splinternet

China approached its hosting of the 2008 summer Olympics as a means of showcasing the country's infrastructure development. The XXIX Olympiad planning team envisioned Beijing greeting the world as a sophisticated metropolis offering the amenities of any other destination city, including significant upgrades to internet, architecture, transportation, and high-end shopping opportunities.

This extraordinary public relations effort was meant to signal to the world that China was open for business, with notable effort made by the government to quell human rights and pollution concerns.

While China has continued to build a robust internet infrastructure in its urban centers, the Chinese Communist Party also blocks most Western digital content,[23] effectively eliminating a degree of foreign influence and increasing its own control over messaging.

The internet has been filtered in China to one degree or another since 2000 via the Golden Shield Project, also known as the Great Firewall of China.

That surveillance has increased markedly in recent years, with the Chinese government increasingly blocking external content, banning Chinese citizens from using certain words and images, and utilizing artificial intelligence to monitor the internet, locate individuals and quell the potential for political unrest.

---

21 "Indonesia urges tech platforms to sign up to new licensing rules or risk being blocked" by Kate Lamb & Stanley Widianto, Reuters.

22 Nor are they necessarily shared from American to American or state to state. But there are court rulings, laws, and constitutional protections that set precedent for various freedoms in the United States.

23 Curious what's blocked and what's not? Go to https://www.comparitech.com/privacy-security-tools/blockedinchina/ and put in a URL. At this writing, CNN is available in China, but Fox News, *The New York Times*, The *Washington Post*, Facebook, Instagram, and Twitter are all blocked.

Yaqiu Wang, a Chinese researcher at Human Rights Watch, described the shift in China's approach to the internet this way:

> Against the backdrop of China's economic rise and growing influence around the world, the Communist Party has been ramping up its nationalistic propaganda, promoting the idea that a diminishing West, especially the United States, is determined to thwart China's rise.
>
> When so few have alternative sources of information, government propaganda becomes more believable: The coronavirus was brought to China by the U.S. Army; protesters in Hong Kong are "violent and extreme" and instigated by U.S. intelligence; the election of pro-independence candidate Tsai Ing-wen to Taiwan's presidency was a result of American manipulation.
>
> Inside China, people are living in an information bubble that the government is getting better at controlling.[24]

In 2015, China launched "Made in China 2025," a 10-year plan aimed at securing China's dominance of global technology. Part of that strategy was for China to work with private investors to buy technology companies in other countries. Investment in the U.S. was significant, particularly with innovative startup firms that focused on things like artificial intelligence, self-driving vehicles, augmented and virtual reality, and space travel.[25]

A 2018 Pentagon report concluded that Chinese investments in emerging technology were enabling a "strategic competitor to access the crown jewels of U.S. innovation." A key point of the report was "the U.S. does not have a comprehensive policy or the tools to address this massive technology transfer to China."[26]

Another policy aimed at digital domination is that of **Golden Share**, which allows the Chinese Communist Party to take partial ownership of private Chinese companies. The party is particularly interested in companies that collect large amounts of user data and disseminate information. For example, in 2019 the government claimed one of only three seats on the board of directors of ByteDance Technology, which owns both Douyin in China and its popular global counterpart, TikTok. China also acquired a 1% ownership stake in the company.

Other countries that have stepped up infrastructure development and information control in the past decade include Russia, which has flirted with creating an internal internet, RuNet, that could digitally isolate Russia from the rest of the world.

Through what's loosely known as its sovereign internet law, Russia requires all web and internet hosting services to be on Russian soil. This centralized infrastructure control allows the Russian government to use the internet for surveillance of its citizens, as well as isolate the country from the global internet, as desired.

The Pandemic's Influence

Researchers have found the pandemic sparked a global increase in the filtering and blocking of information by platform and internet service providers in democracies, monarchies, and autocracies. Some of this was due to countries passing restrictive new policies, which put the requirement for censorship on the providers.

---

24 "In China, the 'Great Firewall' is changing a generation" by Yaqiu Wang, *Politico*.
25 "How China acquires 'the crown jewels' of U.S. Technology" by Cory Bennett & Bryan Bender, *Politico*.
26 "China's technology transfer strategy" by Michael Brown & Pavneet Singh, Defense Innovation Unit Experimental.

Freedom House, an organization that evaluates and ranks internet freedom worldwide, pointed to the fact that, "in the Covid-19 era, connectivity is not a convenience, but a necessity."[27] Nonetheless, researchers determined that, in the first year of the pandemic,

- 13 countries experienced government-imposed internet shutdowns during the pandemic,
- 20 countries expanded laws restricting online speech,
- 28 countries censored websites and/or social media, and
- 45 countries arrested or detained internet users for Covid-19 related posts.[28]

Freedom House also found countries like China and Russia are engaged in a race for "cyber sovereignty," imposing regulations that restrict the flow of information across borders. In short, many countries have used the pandemic as an excuse to increase competitive advantage and shutdown the potential for dissent.

It's notable that the U.S.—traditionally one of the leaders in promoting free speech on the internet—posted four consecutive years of declining internet freedom (2017–2020), due largely to government policies that target technology created by and information flowing from other countries.[29]

University of Michigan researcher Ram Sundara Raman concluded that the uptick in censorship and regulation effectively fragments the global internet, increasing the global divide:

> We imagine the internet as a global medium where anyone can access any resource, and it's supposed to make communication easier, especially across international borders. We find that if this upward trend of increasing censorship continues, that won't be true anymore. We fear this could lead to a future where every country has a completely different view of the internet.[30]

Other concerns focus on ways the internet and, by extension, information are increasingly vulnerable to disruption by digital attacks from a host of individuals and government-sponsored hackers.

Digital Disruption: Hacktivism, Cyberterrorism, & Cyberwarfare

A notable side effect of our reliance on digital communication technology is that when it fails in a big way, we're all about one step away from chaos. There are those who tap into this disruption to spur change, those who cause it because they're paid to (or just because they can), and those who have discovered the benefits of cyberwarfare.

In short, cyber attacks are pretty much an unavoidable part of digital living. Most attacks are the work of **hackers**, people adept at understanding both the complexities of computers and networks and the potential vulnerabilities to disrupt them.

Hackers generally are trying to gain illegal entry into a system for some type of gain: change, money, or influence.

Mark Zuckerberg has embraced the hacker ethos since the founding of Facebook, including his famous motto of "move fast and break things." He says the term hacker is misunderstood:

> In reality, hacking just means building something quickly or testing the boundaries of what can be done. Like most things, it can be used for good or bad, but the vast majority of hackers I've met tend to be idealistic people who want to have a positive impact on the world.[31]

---

27   "The pandemic's digital shadow" by Adrian Shabaz & Allie Funk, Freedom House.
28   "Information isolation: Censoring the Covid-19 outbreak" by Adrian Shahbaz and Allie Funk, Freedom House.
29   "The pandemic's digital shadow."
30   "'Extremely aggressive' internet censorship spreads in the world's democracies" by Nicole Casal Moore, *Michigan News*.
31   "Founder's letter, 2012" by Mark Zuckerberg.

Nonetheless, hackers are blamed for digital disruption every day across the globe.

One of the most common hacks is a **Distributed Denial of Service** (DDoS) attack, which involves multiple compromised systems—often thousands of individual computers—being used to target a single system. If you send enough requests to open just about any single website at once, it is likely to crash.[32]

Here's the basic idea behind a DDoS attack:

- A hacker sends a virus disguised as a link or attachment in an email or text.

- One click and a Trojan virus—something that appears to be something it's not—is planted on your computer. New viruses emerge all the time, so if you're not updating your anti-virus software, or don't have any, you're more vulnerable to these types of attacks.

- From there, that sleeping virus waits until it's activated by a hacker.

- Your computer, now part of a zombie cyber army, is directed to send messages to a single site, overloading and crashing it.

A notable DDoS attack took place in fall of 2016, when hackers brought down servers at Dyn, one of many companies that monitor parts of the internet's routing network, the Domain Name System.[33]

Dyn was attacked three times in one day, with tens of millions of internet-connected devices sending messages to the company's servers.[34] The outage, which lasted for hours, also slowed or stopped traffic to websites like Netflix, Reddit, Spotify, and Twitter.

In a then-novel twist, the **internet of things**—the near-ubiquitous connection of just about any electronic device to the internet—enhanced the success of the Dyn attack, as networked electronics like televisions, cameras, and baby monitors were also be harnessed to repeatedly contact Dyn servers.

Another common attack, **ransomware**, also involves you clicking on something you shouldn't. Ransomware attacks have been rising in size and impact over the past decade. Thousands of government agencies across the globe have been the victims of ransomware attacks, as well as health care, education, and emergency systems.

Here's the basic idea behind a ransomware attack:

- This type of hacker needs you to enter your organization's credentials and password through some ruse—like a seemingly legitimate problem with your work email account.

- Depending on your level of access to the organization's internal computer network and servers, you've just allowed someone to enter what's otherwise a secure network.

- Once a hacker gets access to parts of the network or a critical database, they digitally scramble and/or lock it, preventing organizational access.

- Then they ask for a ransom, usually cryptocurrency, to unlock the system. Don't want to pay? Your system or database may get deleted or left scrambled beyond repair.

---

32 Kind of like when a new restaurant opens and everyone has to go there on day one. It's never going to end well, as the result is an overwhelmed staff and overburdened kitchen.

33 As we noted in Chapter 9, the DNS is widely recognized as being the least secure part of the internet and therefore among the most likely to be exploited by hackers.

34 "Hackers used new weapons to disrupt major websites across U.S." by Nicole Perlroth, *The New York Times*.

The shift from paper to digital records has facilitated the rise in ransomware attacks. Cyber security expert Avi Rubin explains the impact of that shift like this:

> Imagine if somebody would sneak into a government building at night, load up a bunch of boxes with all the paperwork for all the pending permits and all the pending house closings and all the pending business that the city was conducting, put it all in a truck and drive away—and demand some money in order to bring that truck back. That's a lot easier to do in cyberspace without getting caught.[35]

And that's exactly what happened in 2019 to the city government of Baltimore, Maryland. It was hit by an attack that prevented city workers from accessing systems that allowed residents to pay their water bills, property taxes, and parking tickets. City email was also frozen, as well as the system that allowed real estate transactions to take place.

The city refused to pay the 13-bitcoin ransom (just over $75,000) and spent months restoring the systems. The ultimate cost to the city was roughly $10 million in "cyber-attack remediation and hardening of the environment," as well as an estimated $8 million in lost or delayed revenue.[36]

Who's Responsible for Cyberattacks?

There are three general groupings of hackers who plan and execute cyberattacks: hacktivists, cyberterrorists, and state-contracted agents of cyberwarfare.

Hacktivists—hacker-activists—take over websites, find system vulnerabilities, or sweep data and share it online. The goal generally is social change or attention to a cause and hacktivists often work together.

A popular 2015 television series, "Mr. Robot," featured an Anonymous-like collective that successfully hacked into the fictitious Evil Corp and erased millions of dollars in consumer debt. That fictional effort, spurred by the desire of one hacker—a cybersecurity expert by day—to rescue a childhood friend from crushing student loan debt, nearly collapsed the international financial system.

The actual Anonymous[37] also gathers like-minded hackers around particular causes, sometimes global ones. Following the invasion of Ukraine by Russia in February of 2022, Anonymous used Twitter to declare cyber war against the Russian government.

The collective took credit for attacks such as taking down the government's primary website, leaking the database of the ministry of economic development, intercepting Russian military communications, and taking control of Russian state TV channels to broadcast footage from the invasion of Ukraine.

Cyberterrorists use the same tools as hacktivists to do many of the same things. They may be hackers-for-hire or hackers-for-fun, but the goals are generally disruption and/or profit. And cyberterrorists may work alone, but collectives also take credit for big hacking events, particularly ransomware.

In 2015, a hacker collective took credit for knocking *New York* magazine offline for nearly 12 hours with a DDoS attack. Why? Largely because they could.[38] Perhaps coincidentally, it was the day NYM published its exposé featuring 35 women who claimed to have been sexually assaulted by actor Bill Cosby.

---

35   "Ransomware cyberattacks knock Baltimore's city services offline" by Emily Sullivan, NPR.
36   So, why didn't they just pay? It's complicated, but most cities follow FBI guidance that recommends not paying terrorists. *The Baltimore Sun* covered the May 2019 attack and its aftermath extensively, as well as a 2018 attack on the city's 911 system and a 2020 attack on the school district that kept students from online instruction during the early pandemic. Search for the stories online.
37   Anonymous is covered in more detail in Chapter 7.
38   Although a hacker who took credit for the attack, ThreatKing, said it was because they had a bad experience in New York and wanted to punish anything with the city's name in it.

Researcher Molly Sauter provided a physical-world analogy, as well as noting the conflicting goals of the hacker and magazine:

> This is like someone buying out the entire print run of *New York* magazine, for the same reason—that they didn't want anyone to see the cover. Unfortunately, this is a case where some hacker wanted to do something, and unfortunately it coincided with a day when New York magazine wanted to do something bigger.[39]

And then there was the guy who single-handedly took down every website operating in North Korea in retaliation for being hacked by what he said were state-sponsored spies. The hacker, who goes by the handle P4x, was disappointed that the U.S. government hadn't responded to foreign hacks against its citizens. So, he went after North Korea on his own—because he could. "I want them to understand that if you come at us, it means some of your infrastructure is going down for a while," he said.[40]

Had U.S. government hackers retaliated, rather than an individual, the attack would have amounted to an increasingly common form of warfare between countries.

Cyberwarfare is state-sponsored hacking—nations engaging in digital attacks on other nations. These hackers may use the same, or far more advanced, tools as hacktivists and cyberterrorists, with goals of disruption, retaliation, or policy change. Often the hackers are loosely coupled to the government, which allows the country a degree of plausible deniability (it wasn't us).[41]

U.S. cyber security teams have been active participants in these types of attacks, launching one of the earliest in about 2008. Stuxnet, a destructive malware attack on an Iranian nuclear facility was attributed to a partnership between the U.S. and Israel—aimed at preventing Iran from creating nuclear weapons. The digital worm[42] was discovered in 2010 as scientists tried to figure out why Iranian systems kept crashing. At that point, it likely had been wreaking havoc on the system for at least a year.

Cyberwarfare attacks frequently are denied by governments, both those committing the original act and those who retaliate against it.

A 2014 internet blackout in North Korea is particularly interesting, in that its largely denied origin likely was a cyberattack by the U.S. in response to a largely denied cyberattack on a global film distributor.

A hack of Sony Pictures Entertainment—purportedly by North Korea—released a trove of largely unflattering emails and confidential data onto the internet over the course of several weeks in late 2014. Hackers demanded that Sony not release "The Interview," a comedy that featured James Franco and Seth Rogen hatching a plot to assassinate North Korea Supreme Leader Kim Jong Un. They also threatened attacks on the American public, should the film be released. Sony decided not to release the film.

In late December of 2014, President Obama chastised Sony's decision and condemned the attack on Sony, noting that the U.S. would "respond proportionately … in a place and time and manner we choose." Three days later internet service across North Korea went down for just short of 10 hours as result of a massive series of DDoS attacks. The day after that, Sony released "The Interview."

The origins of both attacks are murky, but most examples of cyberwarfare are. And there are a lot of them.

More recent attacks attributed to cyberwarfare include the Sunburst hack, which was facilitated on malware built into a SolarWinds software update. The hack tapped into critical systems in the technology sector, as well as dozens of federal and state agencies, including the U.S. Department of Homeland Security, National Institutes of Health, and Federal Aviation Administration.

---

39   "So a hacker brought down your news site. Now what?" by Madeline Welsh, Nieman Lab.
40   "North Korea hacked him. So he took down its internet" by Andy Greenberg, *Wired*.
41   "What is cyberwar? Everything you need to know about the frightening future of digital conflict" by Steve Ranger, ZDNet.
42   A virus and worm may have similar effects, but are different things. A virus requires external activation to start the desired action, while a worm is programmed to begin work on its own.

Sunburst is notable for several reasons, not the least of which being that it began in September 2019, but wasn't detected until December 2020. The attack was attributed to an independent hacker collective directed by Russian intelligence service, with the U.S. government imposing economic sanctions on Russia in response.

Russia is more directly accused of mounting a massive malware attack on Ukraine, hours before Russian troops entered the neighboring country. The attack, dubbed FoxBlade, was detected and disabled by Microsoft's Threat Intelligence Center. The organization, a public-private partnership aimed at stopping disruptive cyberattacks, notified the Ukrainian government and offered advice for neutralizing the threat.

Other nation-to-nation attacks are better categorized as **information warfare**, in that they are effectively propaganda campaigns aimed at influencing opinion and changing behavior. For example, "fake news" from Russian "troll factories" were implicated in misinformation and disinformation targeting American voters before the 2016 presidential election.

International Treaties Need to Catch Up

After World War II the Geneva Convention set up rules for warfare, largely to protect civilians. Cyberattacks and an endless flood of propaganda have the potential to seriously damage a nation's infrastructure and/or destabilize its government, affecting average citizens. Neither is addressed in any formal way on the global stage.

This argument has been made since at least 2017, when Microsoft President Brad Smith spelled out the urgent need for a **Digital Geneva Convention** at an annual cybersecurity conference. He noted,

> We suddenly find ourselves living in a world where nothing seems off limits to nation-state attacks. Conflicts between nations are no longer confined to the ground, sea and air, as cyberspace has become a potential new and global battleground. There are increasing risks of governments attempting to exploit or even weaponize software to achieve national security objectives, and governmental investments in cyber offense are continuing to grow. [43]

Smith acknowledged that a global agreement could take decades and likely would come only in the aftermath of a particularly destructive attack.

In the meantime, examples of large-scale cyberattacks with significant consequences to average Americans are everywhere. In early 2021, a ransomware attack forced Colonial Pipeline to shut down its distribution system, which delivered gasoline to much of the eastern and southeastern U.S., for five days. The result was a sudden shortage of gasoline, diesel, and jet fuel that affected Americans living in 17 states and the District of Columbia.

And, Americans did what Americans do, which was to "panic buy"—filling as many vehicles and fuel containers as possible, contributing to the fuel shortage.[44]

The situation underscored the importance of the first amendment of Smith's proposed Digital Geneva Convention: No targeting of tech companies, private sector, or critical infrastructure.[45]

The Colonial Pipeline attack was attributed to a Russian-based hacker collective known as DarkSide, with no known connection to the Russian government. Nonetheless, President Biden accused Russia of harboring and nurturing hacker networks, and called for "a cybersecurity arrangement that begins to bring some order." Russian President Putin in turn accused the U.S. of being the largest source of cyberattacks around the world.

---

43 "Why we urgently need a Digital Geneva Convention" by Brad Smith, World Economic Forum.
44 We also did this with toilet paper, paper towels, and hand sanitizer in the first year of the pandemic …
45 Smith (2017), op. cit.

In short, business as usual, even as cybersecurity experts renewed arguments that international cooperation—not finger pointing—is the only way to thwart damaging cyberattacks and devastating cyberwarfare.

Summary

Internet connectivity is, at best, uneven across the planet. There are a number of proposed solutions to bridging the global digital divide, none of which have proven to be simple or readily accepted. The digital divide is widened by the interests of countries, with individual leaders and governments making decisions regarding development and management of the information infrastructure and how the flow of information is controlled. As a result, there is intense competition among a few countries for control of digital distribution and content. There's also a thriving cyberverse of hackers—hacktivists, cyber-terrorists, and state-sponsored cyberwarriors—who use a range of tools to meet a variety of ends. At present there is tacit agreement that a global solution is necessary, but limited progress in that direction.

Key concepts

Cyberterrorists
Cyberwarfare
Digital Geneva Convention
Distributed Denial of Service
Global digital divide
Golden Share
Hackers
Hacktivists
Information control
Information warfare
Infrastructure development
Internet of things
Kessler Syndrome
Ransomware

References

Bennett, C. & Bender, B. (2018, May 22). How China acquires "the crown jewels" of U.S. Technology. *Politico*. Retrieved from https://www.politico.com/story/2018/05/22/china-us-tech-companies-cfius-572413

Bishop, T. (2022, March 25). Elon Musk, are you out there? SpaceX Starlink internet service gets more reliable. *GeekWire*. Retrieved from https://www.geekwire.com/2022/elon-musk-are-you-out-there-spacex-starlink-internet-service-gets-more-reliable-and-mobile/

Boyle, A. (2022, April 20). NASA awards millions to SpaceX and Project Kuiper for satellite communications. *GeekWire*. Retrieved from https://www.geekwire.com/2022/nasa-awards-millions-to-spacex-and-amazons-project-kuiper-for-satellite-communications/

Brown, M. & Singh, P. (2018). China's technology transfer strategy. *Defense Innovation Unit Experimental*. Retrieved from https://nationalsecurity.gmu.edu/wp-content/uploads/2020/02/DIUX-China-Tech-Transfer-Study-Selected-Readings.pdf

Casal Moore, N. (2020, November 17). "Extremely aggressive" internet censorship spreads in the world's democracies. *Michigan News*. Retrieved from https://news.umich.edu/extremely-aggressive-internet-censorship-spreads-in-the-worlds-democracies/

Clark, S. (2022, May 13). SpaceX passes 2,500 satellites launched for Starlink internet network. *Spaceflight Now*. Retrieved from https://spaceflightnow.com/2022/05/13/spacex-passes-2500-satellites-launched-for-companys-starlink-network/

Duffy, K. (2021, May 29). A SpaceX fanatic created a website to find out when Starlink satellites were visible in his location. *Insider*. Retrieved from https://www.businessinsider.com/how-to-use-find-starlink-satellites-spacex-website-elon-musk-2021-5

Geomagnetic storm and recently deployed Starlink satellites. (2022, February 8). *SpaceX Updates*. Retrieved from https://www.spacex.com/updates/

Greenberg, A. (2022, February 2). North Korea hacked him. So he took down its internet. *Wired*. Retrieved from https://www.wired.com/story/north-korea-hacker-internet-outage/

Hempel, J. (2018, May 17). What happened to Facebook's grand plan to wire the world? *Wired*. Retrieved from https://www.wired.com/story/what-happened-to-facebooks-grand-plan-to-wire-the-world/

Lamb, K. & Widianto, S. (2022, July 18). *Indonesia urges tech platforms to sign up to new licensing rules or risk being blocked*. Reuters. Retrieved from https://www.reuters.com/technology/indonesia-urges-tech-platforms-sign-up-new-licensing-rules-or-risk-being-blocked-2022-07-18/

Mann, A., Pultarova, T., & Howell, E. (2022, April 14). SpaceX Starlink internet: Costs, collision risks and how it works. *Space.com*. Retrieved from https://www.space.com/spacex-starlink-satellites.html

Mattise, N. (2018, February 18). Project Loon team gave Puerto Rico connectivity. *Ars Technica*. Retrieved from https://arstechnica.com/science/2018/02/project-loon-engineer-sees-a-tool-for-future-disaster-response-in-puerto-rico/

Perlroth, N. (2016, October 21). Hackers used new weapons to disrupt major websites across U.S. *The New York Times*. Retrieved from https://www.nytimes.com/2016/10/22/business/internet-problems-attack.html

Ranger, S. (2018, December 4). What is cyberwar? Everything you need to know about the frightening future of digital conflict. *ZDNet*. Retrieved from https://www.zdnet.com/article/cyberwar-a-guide-to-the-frightening-future-of-online-conflict/

Shabaz, A. & Funk, A. (2020a). *Information isolation: Censoring the Covid-19 outbreak*. Freedom House. Retrieved from https://freedomhouse.org/report/report-sub-page/2020/information-isolation-censoring-covid-19-outbreak

Shabaz, A. & Funk, A. (2020b). *The pandemic's digital shadow*. Freedom House. Retrieved from https://freedomhouse.org/report/freedom-net/2020/pandemics-digital-shadow

Smith, B. (2017, February 14). Why we urgently need a Digital Geneva Convention. *RSA Conference Proceedings*. Retrieved from https://blogs.microsoft.com/on-the-issues/2017/02/14/need-digital-geneva-convention/

Sullivan, E. (2019, May 21). Ransomware cyberattacks knock Baltimore's city services offline. *NPR*. Retrieved from https://www.npr.org/2019/05/21/725118702/ransomware-cyberattacks-on-baltimore-put-city-services-offline

Sutter, P. (2021, October 6). Megaconstellations could destroy astronomy and there's no easy fix. *Space.com*. Retrieved from https://www.space.com/megaconstellations-could-destroy-astronomy-no-easy-fix

Wadhwa, V. & Salkever, A. (2022, May 4). How Elon Musk's Starlink go battle-tested in Ukraine. *Foreign Policy*. Retrieved from https://foreignpolicy.com/2022/05/04/starlink-ukraine-elon-musk-satellite-internet-broadband-drones/

Wang, Y. (2020, September 1). In China, the "Great Firewall" is changing a generation. *Politico*. Retrieved from https://www.politico.com/news/magazine/2020/09/01/china-great-firewall-generation-405385

Welsh, M. (2015, July 27). So a hacker brought down your news site. Now what? *Nieman Lab*. Retrieved from https://www.niemanlab.org/2015/07/so-a-hacker-brought-down-your-news-site-now-what-new-york-magazine-found-out-today/

Zuckerberg, M. (2012). Founder's letter, 2012. *Facebook*. Retrieved from https://m.facebook.com/nt/screen/?params=%7B%22note_id%22%3A261129471966151%7D&path=%2Fnotes%2Fnote%2F&refsrc=deprecated&_rdr

CHAPTER 14

Remixes & Mashups: Appropriation of Culture Goes Digital

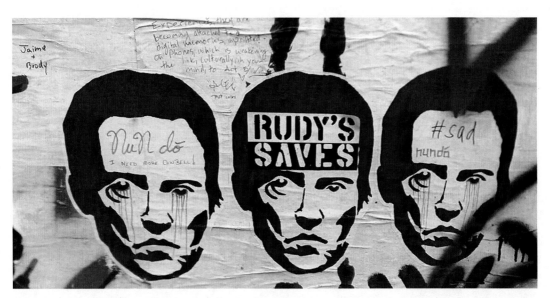

Figure 14.1. Christopher Walken mash-up on the free-form, free-expression wall below Seattle's Public Market. While public spaces like this are rare, the web offers plentiful opportunities for collaboration. (Photo by Susan Wiesinger)

Takeaway: With the combination of powerful, relatively inexpensive home computing and plentiful digital media offerings, everything old can be new again, and everything new can be newer.

———————————

When Julian was in second grade, he proudly presented his mom with a picture he'd drawn at school.

"What a nice dinosaur," she said.

"It's a *dog*," he responded.

"Yes, it's a lovely dog," she said, agreeably.

"Will you put it on the refrigerator?" he asked, hopefully.

"Oh, honey," she said. "This painting is so good that, well, maybe we *shouldn't* put it on the refrigerator."

A short time later, Julian retrieved his dog drawing from the trash. While the original image was lost along the way, Julian held onto the feeling of "no artistic talent" well into his college years.

One day he had an epiphany: He could share his dog on his own digital refrigerator—his blog on the World Wide Web—for all to admire. That would take away the lingering need for validation from his mom, because he could share his "unique blend of contemporary abstract impressionism, raw emotion and inner-angst" with the world.[1]

And, what's more, Julian could allow others to use, reproduce, mock, modify, and almost infinitely share it. This reflects *intercreativity,* which was foundational in the original vision of the World Wide Web. Tim Berners-Lee, inventor of the web, described this as people-to-people collaboration—people co-laboring to create something together. In practice, the web has become a space of remixes, mash-ups, samples, spoofs, knockoffs, and satire—free expression in all its complex forms.

Culture Clash

In the highly mediated digital world, we are swimming in a sea of ideas, stories, arguments, and artistic expressions. In the past, these were the activities of a more select group of people who had the money or the power or the technical and artistic ability to stand above the crowd.

If you wanted to participate in the past, it took a commitment of time and resources that some could afford out of privilege, others because of opportunity (right place, right time, right family connections, right network of friends, etc.).

But since the tools have changed, the ability to participate has as well. The commitment is less costly—just about anyone with access to a computer and/or smartphone can participate for work, as a hobby, or simply as a form of mediated distraction. The cost of good quality still and video cameras, as well as digital audio technology continues to go down—and smartphone tech is remarkably good. But that doesn't mean everyone uses these tools, or uses them well.

It's something of a cliché, but this anything-goes attitude has to do with the idea of participating as an "amateur." The French root of this word emphasizes that one is pursuing an activity out of love. Of course, the term has both positive and negative sides. But when you view the current climate of digital interaction in historical context, you essentially see the battle between "amateur" and "professional" has played out over and over again.

Whether we are talking about journalism or cooking, music or storytelling, the era of elites and experts has been challenged by the wide world of people who get the itch to participate and alter the way things are played.

---

1 Wonder if Julian knows this is still up? See http://julianwilliamson.wordpress.com (2008).

You can see how participation by amateurs might result in backlash from professionals. Questioning the value of authorities and experts means that the justification of their expertise is in question. At one point in history, this was tied up in the way things were written out. Keep in mind that writing is a *storage technology* and that some of that storage technology was being used to contain the right way to do things like grow food, navigate the local seas, and invoke gods through the arts.

But the language of storage was complex, often becoming pictures or other symbols that required a kind of learning that was very exclusive. The elite, who were educated to control this knowledge, were kind of like early programmers in a highly symbolic language. Their ability to "code" and "decode" gave them status.

The Latin alphabet, which became the English alphabet (A to Z), did to that class of exclusive experts what the digital realm is doing to all sorts of categories of expertise. So changes to the communication system level the playing field and broaden out who can get in the game.

In an interesting twist, our digital world of killer apps and social media is underpinned by exceedingly complex algorithms, information architecture, and engineering. Those who can think big and write code have recreated a kind of priestly class that keeps the rest of us in their thrall (think Alphabet, Apple, Facebook).

We are in a position where we can interact and create with the digital tools without necessarily being a programmer. But at the same time we need to see the limitations for what they are.

Sometimes we can know how to use the tools without realizing that, at the same time, the tools are using us. Every communication form, after all, has a set of rules and procedures for how they are used.

We can think of these rules as the Program. Maybe the Program is how the financial system works, or maybe the Program is how you use Microsoft Word to write a paper. As media theorist Douglas Rushkoff argued, we can be the people creating the rules, or we can be the people subject to them; as he put it, we will either "**program or be programmed**."

Rushkoff doesn't mean that you have to become a computer programmer, but that you need to understand what it means to consume programs created by others without question:

> The difference between a computer programmer and a user is much less like that between a mechanic and a driver than it is like the difference between a driver and a passenger. If you choose to be a passenger, then you must trust that your driver is taking you where you want to go. ... The more you live like that, the more dependent on the driver you become, and the more tempting it is for the driver to exploit his advantage.[2]

The earliest books produced listeners rather than readers and writers; the printing press produced elite publishing houses and led to the exclusivity of "mass" media. And devices, apps, and websites produce content contributors and consumers, which give power and influence to the companies that control the technological platform.

For example, think about your favorite social media account: You may choose what to post, but it must be done within the parameters of the platform according to rules set by those in control of the programming.

The Evolution of Culture

To unpack this further and apply it to digital information literacy, we need greater understanding of cultivation, curation, appropriation, and culture.

Cultivation is the idea that artists must be nurtured over time to produce great works of art.

Curation involves the selection, preservation, and display of art from cultivated artists.

Appropriation is how the public responds to the art and artist, adopting and adapting bits into daily life.

2   Rushkoff (2011), p. 9.

Culture classically has referred to enduring parts of any given society that ideally outlive it. Culture typically is not something that can be appreciated in the present; it must be viewed from the future to determine its worth. The works reflect something about the time in which they were created.

Culture is a slippery concept, however, and, as with so much else, has changed markedly with each innovation in information communication technology. The evolution can be roughly mapped as progression through a series of changes in what is defined as culture.

Classic high culture appreciates in value and social meaning across wide swaths of time. The original intent of the artist may be lost, but the artifact—poetry, sculpture, painting, music, dance— remains. And the creators are often hailed as geniuses.

High culture requires:

- Patience and a crystal ball. Talent is nurtured over time and true worth may not appear for generations.
    - Examples: Michelangelo and Rodin, Beethoven and Mozart, Shakespeare and Chekhov, Tolstoy and Hemingway, Caruso and Pavarotti, Baryshnikov and Nureyev.

- *Cultivation* by experts in a particular cultural form.

- Intensive *curation* via exhibits, displays, performances.

- Wealthy, elite members of society who pay living expenses for artists, in exchange for social currency in the arts community. This was referred to as a patronage system.
    - The artists are then allowed to develop and focus on their work.

- Protection, preservation, and limited reproduction of works.

- Reduction of the cultural message to one artist, few symbols.

- *Appropriation* through reproductions.

Generational mass culture segments society along generational lines. What is enjoyed and appreciated by your parents (aka the "oldens") may be laughable to you. The artist tends to be contemporary and provides interpretation of the message.

Mass culture requires:

- Value that emerges over the course of a few years. Artists are nurtured as long as they show potential and remain productive.
    - Examples: international supergroups like The Beatles, The Rolling Stones, The Who; individuals like Elvis, Picasso, Fred Astaire.

- *Cultivation* by corporations.

- *Curation* by individuals, who select and collect mass-distributed works, either physical or digital.

- A voracious mass society that wants accessible entertainment, rather than highbrow art.
    - These people are more likely to appreciate a concept album in vinyl, like Pink Floyd's *The Wall,* than a visit to an art gallery.

- Marketing to encourage intensive consumption in individual and social settings.

– Expansion of cultural symbols by particular artists.

– *Appropriation* through consumption.

Meme culture results in fragmentation of all culture.[3] Messages are endlessly modified for personal purposes and the origins of a work may not even be known. There is no common beginning, middle, or end; in fact, individualized culture invites participation at any stage of the process. Individualized culture requires:

– Near instant gratification. Value is not monetary and is achieved through collaboration and sharing.
   • Examples: 140-character tweets, retweets, shares, videos of 15 seconds or less, Google Images.

– Limited *cultivation,* largely in the form of followers on social media feeds, or organized through branded corporate identities.

– Digital, highly personalized *curation,* such as with a social media feed.

– Personal appropriation of other people's work for modification, remixing, and mash-up into something of your own.
   • Examples: playlists that tell a particular story; internet memes.

– Continual reduction and expansion of cultural symbols.

– *Appropriation* through modification.

The designation of three different types of culture is not intended to imply that the three don't coexist. They do. As with any other "innovation," however, some things recede as others emerge.

Clearly you can still go to the Sistine Chapel and see Michelangelo's curated work. But you also can go to the souvenir shop and buy a reproduction of the art on a postcard, calendar, T-shirt, luggage tag, and so on. That's *appropriation.* You also can have an album collection and share your parents' love for 1970s music. That's *curation.*

The rapid shift to individualized culture is worthy of greater focus because of its potential for individual consequences. The digital world allows us to use programs, apps, and web-based services for pretty much whatever we want. But, if we don't understand the rules that govern the digital realm, we will not play it; it will play us.

Lost in Translation: When Mass and Meme Cultures Collide

In late 2013 Beyoncé made an unexpected, online-only release of her fifth album. Critics scrambled to listen to the 14 tracks and post online reviews in a hyper-competitive media environment. Kitty Pryde, a then 20-year-old musician, liveblogged her review for noisey, a music site owned by Vice.com.

The generational-mass-culture-versus-meme-culture clash emerged in Pryde's review of Track 3, a song called "Drunk in Love."

---

3   A meme is a video clip, image, hoax, joke, whatever, that goes viral—spreading from person to person.

Here's part of the review:

> My Jay Z senses are tingling … pretty sure he's about to rap …
> WAAAAAIT HE JUST SAID "I EAT THE CAKE, ANIME."
> JAY Z JUST F-ING USED THE WORD "ANIME" AS A PUNCHLINE.
> This IS THE BEST SONG I'VE EVER HEARD. [4]

Pryde was correct in her use of the word "punchline" but in a completely wrong way. The liveblog started at 12:41 a.m. and, within hours, a well-known blogger called out Pryde for her "white clueless-ness." Alexander Hardy, who blogs as "the colored boy," took Pryde to task on a number of points and provided several "teachable moments."[5] One was the importance of **cultural context**—looking *beyond you* to understand the influence of cultures and the generations living within them.

So let's take this apart and see where Pryde went wrong, starting with the lyrics. Here's the segment she wrote about:

> Ain't got the time to take draws off, on site
>
> Catch a charge I might, beat the box up like MikeIn '97 I bite, I'm Ike, Turner, turn upBaby no I don't play, now eat the cake, Anna MaeSaid, "Eat the cake, Anna Mae!"

Ah. That's different. There are two significant references in here that highlight the difference between generational mass culture and meme culture. Kitty Pryde is a product of the latter, while the referenced "Mike" and "Anna Mae" are products of the former.

"Mike" refers to Mike Tyson, who for a time was heavyweight-boxing champ of the world. The contextual cues were likely lost on Pryde, who might only know of Mike Tyson for his role in "The Hangover" movie series that started in 2009—when she was 17.

In 1992, however, Tyson was convicted of rape and spent three years in prison. That sheds a little different light on the lyric "beat the box up like Mike." We know it's Mike Tyson Jay-Z is rapping about because the mention of "Mike" is followed by a reference to the 1997 title fight in which Tyson bit off a portion of Evander Holyfield's ear.

The Tyson reference is immediately followed by a lyric about Ike Turner, who was the abusive hus-band of music great Tina Turner. Her birth name was Anna Mae Bullock. Ah. Maybe there's a theme here. So "Eat the cake, Anna Mae" actually refers to a scene in the 1993 biopic *What's Love Got to Do With It,* in which Ike Turner shoves cake in Tina's face and slaps her while they sit in a diner.

Well, that changes things a bit, don't you think? "Eat the cake, anime" is no longer a cosmic joke but instead is wrapped into four lines of lyrics that summon rape and physical abuse of women.

Had Pryde looked beyond her own interpretation of the lyrics, she might have been able to better engage her claim a few songs later that Beyoncé is "getting feminist up in this track." She also might have quickly learned that the song "Flawless," featuring Nigerian author Chimamanda Ngozi Adichie, is indeed a feminist anthem. Instead of consulting The Google, Pryde said, if Adichie "is a Miley Cyrus pseudonym I'm going to bed."

To bring this back around to culture clash, the nature of generational mass culture is that it pro-vides people with common, shared signals that link us together and maintain greater society.

As recently as 1980, it could have been claimed that the United States possessed a semblance of shared culture, facilitated by the mass media. One need only look to the common cultural touchstones embodied by Disney, shopping malls, television shows, big-box stores, and franchise restaurants to

---

4   "Liveblogging the new Beyoncé album with Kitty Pryde" by Kitty Pryde, Noisey, Music by *Vice.*
5   "Eat the cake, anime: On white cluelessness (and Beyoncé)" by Alexander Hardy.

understand the influence of generational mass culture. There is a layer of cultural memory that is learned and passed on—grandparent-to-grandchild, parent-to-child, child-to-peers.

When culture is individualized, information is processed and shared through the lens of one individual's reality. And we're not just picking on Kitty Pryde, as young people have pretty much always thought they knew everything (no offense, the author was once there, too). The point here is that Pryde had the tools literally at her fingertips to fact-check but didn't use them.

There is a contemporary assumption that Gen Z knows everything about information technology because those born within it are digital natives—steeped in digital devices, apps, and the web. People who hire those from this generation expect them to possess some organic degree of digital savvy.

This perception is how 20-year-old Pryde found herself (a) with a great writing gig, and (b) the subject of (fair) attack on the basis of her cultural cluelessness. Digital information literacy is not something anyone is born with; it must be learned and practiced.

And that brings us back to remixes and mash-ups—but first one other small digression is necessary.

Two Ways of Thinking about Cookery

Think about two hypothetical cooking shows. In one, we see a chef go through the painstaking process of doing just about everything herself, from growing the wheat and harvesting the eggs, to hunting and slaughtering the wild boar, to mashing up the tomatoes and garlic, to combining all of these into a Tuscan pasta dish to die for.

On the other cooking show, we see a person opening a can or a jar, unpacking ground meat from the store, boiling premade dried pasta, and so on.

If we compare the two, which do you think would be described as more "authentic"? Our tendency, for many reasons, would be to give greater value to the first, more thorough approach.

"Authenticity" is an idea that has great value and is worth thinking about carefully.

One thing we know about both chefs is that they didn't make up their approaches to creating a pasta dish completely out of nothing. In many cases, a previous cultural tradition or a long apprenticeship formed the foundational knowledge being used, or one could even argue that the expertise expressed through fine quality ingredients reflects a similar kind of collective product.

The significant thing to note is how we organize our expectations of creativity and authenticity. Much of what we find valuable in art is really a complex combination of creating and borrowing.

So when we think of the world of **creativity,** we should think of what we are actually doing. We are being creative, making things up, and being authentic. But this is actually a combination, something between a recipe and an improvisation. We just have to think carefully about what we consider to be authentic creativity versus what raises our suspicions that something is fake, put-on, overproduced, pretentious, and suspect.

We clearly place a higher value on authenticity—on the artist who writes her own songs and sings at live performances (versus remakes and lip-synching); on the person who actually does original work, rather than the person who steals, slaps together, and then claims credit.

So as we think about appropriation, keep in mind the complexity of what we are trying to understand. We might find creativity in sound and images on YouTube worthy of praise—and the ultimate praise of sharing with someone else. Creativity in investment banking and tax avoidance perhaps not so much.

Invention and Arrangement

Think about how you decide to organize and arrange ideas, quotes, and bits of culture. We use elaborate combinations of individual and collective activities to keep track of things. At some points in life it becomes important to put some of those bits together into something creative or argumentative—a five-paragraph paper, a video news story, a mash-up video, an ethnographic portrait, or an autobiography.

The habits of collecting and organizing the bits and pieces change as our activities change, and we are often very unconscious of the organizational scheme—at least until it fails!

Part of arrangement is considering two things: how we know what belongs with what, and how we know *just how* the things we know are organized. There is a combination of intuition and practice with each of these. Learning and remembering in a digitally rich environment is something worthy of more attention.

Some of us just use our memory and our ability to recall. The smartphone that can access the internet helps to turn our ability to recall into our ability to search and organize. We might even think of this as the accumulation of cultural evidence of one kind or another, just waiting for the right argument and the right claim to put that evidence into action.

"Right" in this sense ought to be thought of carefully.

This brings us to Aristotle and something he discusses in his *Rhetoric*. He talks about how we put together persuasive arguments through five canons (rules or basic principles):

Inventio—invention or discovery of arguments
Dispositio—organization or arrangement of arguments
Elocutio—mastery of style in argument
Memoria—discipline of recalling arguments
Pronuntiatio—delivery, voice, and gesture of the argument

These canons have withstood more than 2,000 years and become the common terms of invention, disposition, elocution, memorization, and pronunciation.

Here we want to be concerned with the relationship between invention and disposition, or arrangement. Our usual way of recognizing creativity is to think that individuals create things. But think for a moment about *Inventio* mentioned above.

Invention and discovery are in a juxtaposed position. Think of it like the process of writing a song; it's about *finding* the right lyrics, and *finding* the right chords to get where you are going, and even *finding* the right bits of culture to reference in passing.

So let's set aside the idea of wholesale invention for the moment and consider the importance of **disposition**—the organization and arrangement of the elements. You can see where this is leading: In mash-up and remix culture, we have moved passed the idea of invention (aka creation out of thin air). Instead, we prize the person who can put together original pieces in a new way that works.

Appropriation as an Art Form

Think about how hip-hop constructions are often a woven fabric of allusions, commercial culture repurposed, clichés given a new context, and verbal gestures that find a new way to exist. This is seeing arrangement, appropriation, and mash-up coming together to produce great art, great argument, great digital culture.

But the focus on disposition also can be seen as another way to think about invention.

Listen to a piece of jazz music and think about what the musicians are trying to do. They start with a basic structure, say, 12-bar blues. These musicians know the order; they know the disposition of the chord changes.

So when it comes time for the solo, they know which notes will fit, which will cause tension, and *on the spot* they make musical art out of the decisions they make with those notes, based on their deep and technically profound way of arranging them in ways that respond in the moment. But keep in mind, that creativity comes after untold hours of practice, and going through the repeated process of reacting spontaneously in the moment.

Sometimes jazz artists want to drop in pieces that you would recognize—bits of a familiar tune, a recognizable riff. But sometimes they want to get away from that, to improvise in a clean way unstructured by organizations of notes that come from the past.

This kind of work creates a path through what you already know and what you can make up at the moment. Think of slam poetry, street dancing, DJ'ing, street art, or perhaps other forms that combine spontaneous creation and the arrangement of known bits and pieces.

The ability to string together images and sounds in ways that are totally recognizable and totally fresh is the jazz of the digital world—machinima videos, trailer parodies, cosplay, fan fiction, role-playing video games. Each of these explores a dynamic tension between a preworn path and striking out into the woods.

The possibilities seem wide open, but they are also limited. Frequently we enjoy those limitations as structures to organize our digital world.

Take, for example, a viral chain that asks you to make a list of the 10 most influential books, post it, and link to other people to make a list of their 10 most influential books. It's something of a game when we respond to such online requests, because of the number, the limits, and the structure, and, of course, considering the next person we might want to tap on the shoulder to get that person into the game.

The High Potential for Legal Entanglements

Of necessity this means we need to start considering the idea of intellectual property. Who owns ideas? Who gets to use them? Who gets to profit from their use?

This is a very complex area and one you have probably had to consider when you decide what media to buy, what media to borrow, and what media to copy. Or perhaps you are trying to think about media you might sell one day soon. Maybe you are a YouTube sensation, and you have already been commodified.

Remember that using/borrowing/copying an idea does not really diminish the original in and of itself. But what it does change is its **scarcity,** the value it has because it is rare or unique to a particular space and time.

The more you move media around, the more you might want to learn about cultural rules about things like copyright, intellectual property, authorship, and distribution. You have some rights to do things, and you have to respect the rights of others, just as you expect your rights to be honored.

Copyright Protects Original Works Online

What types of things result in removal of content for copyright violations? It could be just about any image, comic, phrase, and/or media clip found to be an original creation of an individual or company. **Copyright** gives people and organizations the exclusive right to distribute their own content.

Let's say you find a particularly good wildfire photo from a news service online, print it out, paste it on a poster, and write, "Protect Our Forests!" in big, bold letters. You stand on a downtown street corner and wave it at every passing driver. While the copyrighted content on that one handmade poster in the physical world violates copyright, your intended use of the image is limited by both time and space.

Now let's say you create a Facebook page and use the same image to promote the same views. Your image is now on the equivalent of a global street corner, where it could remain in perpetuity. And while your political views may be protected speech, use of the copyrighted image to illustrate them is not.

What happens when user-generated digital content and copyright claims collide? Here are two brief case studies, one with a fiscal cost and the other with an emotional toll.

Use of a Stolen Photo on a Class Project

A student used a photo of Pearl Jam lead singer Eddie Vedder to illustrate a music review included in a university journalism class project. The student had found the photo through a web search and credited it to the photographer. The class project was published on a journalism department website and also linked to the online edition of the university's student-run newspaper.

Within a few months, the photographer discovered the pilfered photo and demanded that it be removed. It was, and the photographer was satisfied with the department's quick response.

Two years later, the university shifted to a new server system. The page with the photo was inadvertently republished, as it had not been deleted from the department's web directory.

The photographer again found the photo and this time was angered by its continued use. He called the department and demanded payment for the photo. In negotiating with the photographer, the department chair refused all demands for monetary compensation and maintained that the use was educational, that the department had not profited from the photo's misappropriation, and that the use did not affect the photographer's ability to make money from the image.

As apology for the error of republishing the photo, the department chair arranged for four weeks of free online advertising for the photographer's wedding business in the student newspaper. The photographer was satisfied and the chair happy to correct the initial student mistake of using a copyrighted photo without explicit permission. This time the photo was deleted from the web directory.

This story has a relatively happy ending, as there was no financial loss to the photographer and no cost to the student newspaper, the department, or the university. Had this situation involved any other type of publication and a less understanding photographer, it could have been an expensive lesson.[6]

Copyright Compromises Free Speech Online

What happens when a web user gets caught between a platform service provider and another corporation in a copyright dispute? This captures it pretty well:

> One day, you are a minor YouTube celebrity with more than 1,850 subscribers to your personal channel and nearly 1 million viewings of your videos logged. The next day—poof!—it's as if you never existed.

That paragraph is the lead in a news story[7] about a YouTube user whose account and channel were removed after he repeatedly posted brief clips from a variety of news and comedy programs to illustrate political points.

The video blogger, whose user name was LiberalViewer, had run afoul of Viacom, which used the Digital Millennium Copyright Act as leverage to require Google/YouTube to remove more than 100,000 video posts that illegally included Viacom content.

An automated search by Google/YouTube found only about one-third of LiberalViewer's 60 brief media clips were from Viacom programs. But because he had been warned about copyright violations before, his entire YouTube channel was removed. LiberalViewer responded with frustration that his rights to free speech had been usurped by a dispute between two megacorporations:

> I greatly respect and enjoy the products of both YouTube and Viacom, and I'm not looking for a confrontation with anyone. I feel like two big corporations are fighting without thinking about how their actions affect the little people like me.

Copyright 101

While the complexities of copyright law are beyond the scope of this book, it is relatively safe to say that if you didn't create it, you need to ask for permission to use it. Here are a few basics to remember:

---

6   Thanks to Professor Emeritus Glen Bleske (personal communication, August 13, 2014) for sharing this story with the authors so that *you* can avoid making the same mistake.

7   "YouTube Purges LiberalViewer's Account" by Sam McManis, *The Sacramento Bee*.

1. **It doesn't have to say it's copyrighted.** By virtue of being published on the web, it is—unless it explicitly says it can be used under a Creative Commons license. And even then there are attribution and use rules that must be followed.

2. **If you're asked to take something down, take it down.** If a person posts copyrighted material on any website and the copyright owner complains, a service provider is bound by the Digital Millennium Copyright Act to ask you to remove the content immediately. If you don't remove it, you can be sued or charged fair market value for use of the copyrighted media for every second it appears on your site.

3. **You are responsible for the content *you* post.** If someone else shares a copyrighted image in comments on your Facebook feed, for example, that person is responsible for any consequences of the copyright violation.

4. **Right clicking is wrong.** Google Images does not exist as a source of free photos for your public use. The search engine is indexing images of a search term. It exists so you can do things like see what a black widow spider *looks like*, not for you to cut and paste a photo onto a public post about the spider you think you just saw in your basement. That's a significant distinction.

5. **Giving credit to the original source doesn't solve the problem.** The nature of the web being what it is, simply crediting the site where you got the image is not enough. You might just be compounding the problem by linking to a blog post that's already violating copyright law by posting the image.

6. **Your content exists on a website at the whim of your host.** If you happen to be a repeat copyright offender, it's within a service provider's rights to remove your account.

Copyright law rewards users who contribute original photos, artwork, videos, music, and writing to the digital conversation but punishes those who steal the contributions of others.

But it is important to keep in mind that the laws that determine "right" and "wrong" are created and influenced by people. Those people determine and change the ideas of what you can and cannot do based on their interests. Their interests may reflect a moral statement about intellectual property, but they may also reflect corporate-driven notions of how the business of creativity should work. Sometimes their interests are the same as yours. Sometimes they are not.

In short, copyright law frequently benefits big companies with big brands to protect and deep pockets to protect them. It offers very little protection to individual users, particularly on social media where sharing is the norm.

It's up to all of us to promote learning about our rights and about the way our rights are constructed, changed, extended, and limited.

Like with many elements of cultural participation, if you want to have any say in how things work, you have to get involved in the cultural conversation about it. And to do that you have to understand the way things are and how they got that way.

That is the fundamental power of information literacy. It allows you to understand not just what you can and cannot do but how it got that way.

It would take another book this size to detail the way copyright and fair use rules got to their current state and another book to discuss how changes in culture and technology make it necessary for us to plan how this will work in the future.

But you do yourself a favor in realizing that change is not stopping anytime soon, so maybe you should just learn where things are and how they got that way as best you can and act in good faith in your own best interests, especially when it comes to a culture of digital media overload, where it's likely to be *you* who gets remixed and mashed up.

Summary

Culture evolves, often through technological advancement. Varying levels of culture remain, however, even as the definition of culture has evolved from original works by nurtured artists to the near instantaneous spreading of works created by one person, or group, that are modified and shared by another. Cultural appropriation, which always has existed to some extent, is accelerated through the use of digital tools that allow us to remix and mash up anything and everything into something new. The basic idea is that we take pieces of content published by and belonging to others and add our own twist to create memes. Memes add to the digital conversation, particularly when a clever mash-up goes viral and is spread person-to-person for millions of individual views. Sometimes the intent of the original artist is lost—or misinterpreted—in translation. Copyright violations are a valid concern online, where Platform Service Providers have the legal power to remove your content if someone claims a copyright violation. In short, your online content exists at the whim of your host.

Key concepts

Appropriation
Authenticity
Classic high culture
Copyright 101
Copyright
Creativity
Cultivation
Culture
Cultural context
Curation
Generational mass culture
Meme culture
Program or be programmed
Scarcity

References

Aristotle. (4th century B.C.). *Rhetoric* (W. R. Roberts, Trans.). Retrieved from https://ebooks.adelaide.edu.au/a/aristotle/a8rh/complete.html

Hardy, A. (2013, December 13). Eat the cake, anime: On white cluelessness (and Beyoncé). *the colored boy*. Retrieved from http://www.thecoloredboy.com/2013/12/eat-cake-anime-on-white-cluelessness.html

McManis, S. (2007, February 8). YouTube purges LiberalViewer's account. *The Sacramento Bee*, p. 39.

Pryde, K. (2013, December 13). Liveblogging the new Beyoncé album with Kitty Pryde. Noisey, Music by *Vice*. Retrieved from http://noisey.vice.com/blog/liveblogging-the-new-beyonce-album-with-kitty-pryde

Rushkoff, D. (2011). *Program or be programmed: Ten commandments for a digital age*. New York, NY: Soft Skull Press.

Williamson, J. (2008, November 7). *Like I needed another thing to confuse and annoy me*. Retrieved from https://julianwilliamson.wordpress.com/2008/11/07/like-i-needed-another-thing-to-confuse-and-annoy-me/

CHAPTER 15

Yours, Not Yours: Digital Surveillance & the Privacy Paradox

Figure 15.1. With our lives playing out in the digital realm, it's no longer government surveillance as a primary threat. Issues of privacy and ownership are more complex than at any other time in history. Pearls Before Swine © 2018 Stephan Pastis. Reprinted by permission of Andrews McMeel Syndication. All rights reserved.

Takeaway: We give our privacy away for free digital services. It's a good deal if you don't mind your every move being tracked in both the digital and physical worlds. That smartphone in your pocket does more than make calls.

There's a scene in the 2019 movie "Booksmart" that offers a bit of contemporary social commentary to consider. Best friends Amy and Molly get into an argument at a party and, as the two intensify their very personal discussion via a single-scene take, there's a subtle shift in the background: the phones come up, the flashes bloom.

As we witness the fight drawing attention at the party, we also can recognize that their friendship-shredding words will soon be online, for anyone and everyone to watch, share, put to music, and memeify toward virality.

That "Booksmart" scene represents only one of the privacy issues that plague today's digital living. While the people holding the devices present a particular concern, as they're *us*, there is a remarkable amount of digital surveillance that we have less control over—which is saying something.

Surveillance was once something that law enforcement did to gather evidence toward prosecution of criminal acts. Today, **digital surveillance** has been folded into our daily lives through the software and digital devices we—and others—use. It's literally everywhere digital devices can connect to the internet, further eroding the concept of privacy in the 21st century.

The concept of digital privacy is a particularly complicated one. With social media companies and other services, such as Google Workspace,[1] the basic idea is that we allow companies access to our data in exchange for free services.

But technologies that monitor everything from your location, to your usage patterns, to your physical well-being also include software, smartphones, computers, and a wide range of internet-connected devices. Let's explore.

Software & App Surveillance

"Software" seems like a word from the past, a time when people bought boxes of disks to install programs onto their computers. But we still use a lot of software. Some of it comes bundled on our devices—native apps designed and/or preferred for a particular device like operating systems, browsers, search engines, and utilities.

You can think about a **native app** as anything that comes pre-loaded on your mobile device or computer. For example, iPhones and iMacs come with iOS as the operating system and Safari as the browser; Pixels with Android and Chrome. Smartphones also generally have proprietary camera, phone, and other apps that you can't readily remove. Most of us also choose to download apps that meet our needs for convenience, information, shopping, and entertainment.[2]

On computers, nearly every website you go to plants a **cookie**—a tiny bit of tracking software—that allows data and advertising companies to follow your path across the web. Some of these do beneficial things, like remember passwords and the contents of a virtual shopping cart. Others are simply following you around, observing what you do, and trying to anticipate what you might buy.

In reality, cookies may soon be a thing of the past, with browsers beginning to eliminate their use due to privacy concerns. Instead, companies are increasingly relying upon your **browser fingerprint**,

---

1   Google Workspace = Gmail, Calendar, Drive (cloud storage), Meet, Docs, Sheets, Slides, Chat, and Sites.
2   We'd recommend at this moment that you take a look at your phone and inventory the native and downloaded apps, but we fear you'd fall into one and end up ordering DoorDash instead of finishing this chapter. No, wait! Don't! Come back …

which is as unique to you as the ones on your hand. With fingerprinting, browsers collect and share information such as your location, keyboard language, operating system, and the path of sites you visit. While cookies can be deleted or blocked, fingerprinting is invisible to the user.[3]

On mobile devices, nearly every app has an embedded **tracker**. When you download an app, you give that tracker permission to monitor a lot of things on your smartphone—including contacts, camera, location, etc. Unlike cookies, app trackers exist purely for collecting data about you, largely to target advertising.

Don't want cookies or trackers? There's virtually no way to opt out, short of not using the technology. And we're not slowing our technology engagement: In 2021, there were nearly half a million app downloads *per minute* globally, with consumers spending nearly 4 trillion hours on their mobile devices.[4]

Every piece of software on your computer or app on your smartphone is reporting back to the mothership on a regular basis. That means even if you're not using the app or your phone, the app can share your data to whichever company opts to pay for it.

Tech journalist Geoffrey Fowler set out to track the trackers and see what data they were harvesting and sending from his phone. Over the course of a week, he found more than 5,400 trackers, that consumed a significant amount of data and battery life—all with no action by him. He discovered that some trackers were sending his exact location, contacts, email address, and unique smartphone ID.[5]

That ID is known as an **Internet Protocol address**, and every digital device that connects to the internet has one.[6] If you use the same set of devices all the time, even if you're rotating among them, all paths lead back to you.

A notable example of app tracking uncovered by Fowler was from the popular DoorDash app, which sent data to nine third-party trackers every time he opened the food delivery app. Fowler explained what the trackers do:

> One tracker called Sift Science gets a fingerprint of your phone (device name, model, ad identifier and memory size) and even accelerometer motion data to help identify fraud. Three more trackers help DoorDash monitor app performance—including one called Segment that routes onward data including your delivery address, name, email and cell carrier.
>
> DoorDash's other five trackers, including Facebook and Google Ad Services, help it understand the effectiveness of its marketing. Their presence means Facebook and Google know every time you open DoorDash.[7]

That points to another issue with cookies and trackers alike: most report to ad service companies, the bulk of which are owned by Meta and Alphabet. If you choose to sign into apps with Facebook or Google, you're giving those companies a great deal more access to your data, so they can micro-target you with advertising. Fowler noted that every bit of *your* data that a cookie or tracker conveys increases your vulnerability in the inevitable event of a data breach.

Apps that have access to your data also have the ability to share them with law enforcement and groups seeking information about individual behavior. This became a concern with period-tracking

---

3   "The quiet way advertisers are tracking your browsing" by Matt Burgess, *Wired*.
4   "The State of Mobile in 2022," data.ai. Look for the updated report online.
5   "It's the middle of the night. Do you know who your iPhone is talking to?" by Geoffrey Fowler, *The Washington Post*.
6   To find the IP address on your smartphone, go to Settings and look for About phone (usually at the bottom) of the long list of options. An IP address has four sets of numbers, separated by periods. Each number is within the range of 0 to 255. For example, 19.0.7.235.
7   Fowler (2019), op. cit.

apps following the U.S. Supreme Court's 2022 reversal of Roe v. Wade, the 1973 decision that deemed abortion a constitutional right.

Period-tracking apps are used by women to monitor their menstrual cycles and track sexual activity, whether for convenience or fertility. As with other apps, period-trackers share data with third-party companies—all linked to the user's identity. That data could then be mined by anyone with the funding and incentive to track down pregnant women, particularly young ones who live in states where abortion is illegal.[8]

Mobile Surveillance

Think of how much of your identity you carry around with you every day in that surveillance device (aka, smartphone) in your pocket. It has ready access to all of your contacts, photos, social media, music, banking, shopping, gaming, and more.

While you may use your face or fingerprint to open your smartphone, that's only protecting you from other people getting their hands on it. The real data sharing takes place by the device and apps—with or without your participation.

We know we rely on our smartphones heavily and use them too much, but we're not really interested in stopping. A recent Gallup survey[9] found that 65% of users said smartphones made their lives better. Nearly half said they couldn't imagine life without their device and would feel anxious if they misplaced it even for a short period of time. More than 7 in 10 adults surveyed said they keep their smartphone near them 24/7. That number spikes up to 9 in 10 for those in the 18–29 age group.

And smartphones aren't the only devices we have on us continually that are collecting large amounts of personal data. How about the one on your wrist that keeps track of your location, sleep patterns, heartrate, breathing, body temperature, and more?

Wearables, like smartwatches and fitness trackers, collect and share a significant amount of health data. On the upside, Fitbit allows users to participate in a study that will notify them if the device detects potentially problematic heart rhythms. On the downside, Fitbit is owned by Alphabet, so all that health data is going to a company that already specializes in collecting and mining your data.

Amazon also has entered the wearable market with Halo, a wrist-placed fitness tracker than does everything a Fitbit can. It also has a couple of added features that reviewers classified as "creepy" and "invasive": a body fat scan and Tone, which listens to your voice all day long and provides feedback on your behavior and mood. The latter feature has been criticized for recording and analyzing the speech of others nearby.[10]

It's always worth asking why Alphabet, Amazon, or any other Big Tech company might be interested in a particular market. And health tracking clearly is a profitable one, ripe for innovation—a fact Amazon doubled down on when it spent nearly $4 billion to acquire One Medical, a successful healthcare provider network. Amazon also owns an online pharmacy and a virtual urgent care service, both of which were created through acquisition of other companies.

Notably, health tracking expands the data points these companies can collect and share. How that volume of data can be secured is a valid question, particularly since no tech company to date has a clean record when it comes to massive data breaches.

Internet of Things Surveillance

The **Internet of Things** refers to the wide range of internet-connected devices that are capable of spying on us day in and day out in our homes, cars, neighborhoods, streets, businesses, public transit,

8   "With Roe overturned, period-tracking apps raise new worries" by Tatum Hunter & Heather Kelly, *The Washington Post*.
9   "Americans have close but wary bond with their smartphone" by Lydia Saad, Gallup.
10   "Amazon Halo review: The Fitbit clone no one asked for" by Victoria Song, *The Verge*.

and workplaces. We may not use them for communication, but they talk to each other and can be harnessed for data gathering—and hacker-driven botnets.

There are two basic types of these devices to consider: internet-connected devices belonging to you and internet-connected devices belonging to others.

Things We Choose

Our homes are filled with internet-connected devices that we've invited in, such as smart TVs, speakers, video game consoles, lights, thermostats, refrigerators, baby and pet monitors, and more. They offer a lot of entertainment, convenience, and security, but also provide a surprisingly public window into your personal world. Virtually all collect and transmit data.

Smart TVs have a relatively long track record for transmitting data about what you're watching back to the companies that make them and to the third-party apps you choose. Think Vizio, Samsung, Netflix, Hulu, etc. The Federal Trade Commission ruled in 2017 that TV companies had to allow consumers to actively choose to be tracked, not opt out later. That said, the decision to opt in is built into the fine print that crosses the screen as you set up a new TV. In short, it's like a user agreement that no one reads.

And TV data gathering and distribution doesn't just cover that device anchored in your living room. There's a reason apps like Netflix want you to tell them who's watching and set up separate profiles—they want to be able to isolate your data. Tech journalist Geoffrey Fowler explains it like this:

> Data firms use your TV history to link up what you watch with what you do on your phone, tablet and laptop—even what you buy in stores. It's as if your TV can unhook itself from the wall and follow you around.[11]

Smart speakers and other voice assistants—think Apple's Siri and Amazon's Alexa— also pose privacy concerns, as they're literally always listening. Unlike TVs, the utility of smart speakers can be a bit more difficult to figure out. But those who do invite them in do so because they're very good for music and podcasts, activating other smart devices (like lights and alarm systems), and answering useful and stupid questions.

One of the biggest concerns people tend to have regarding smart speakers is that they're not all that smart (yet) and require significant human monitoring to fine tune the technology. For example, Amazon has employed thousands of people to listen to recordings made by its Echo device, transcribe them, and feed them back into the system to improve speech recognition and language patterns.[12] Can that data be linked back to you? Sure—each smart speaker has a unique IP address.

Same with gaming consoles. The data they collect can be used by game developers to nail down player demographics, personality traits, habits, and other data than can be used to increase their attention/addiction to gaming.

Ads within games also collect data, in addition to encouraging us to download and play new games—or to pay to avoid ads. For example, if you watch or interact with an ad, you likely will see more ads like that in the near future.

In the purest surveillance sense, gaming consoles and computer monitors also can be used to enforce gaming restrictions. For example, China has classified gaming as "spiritual opium" and imposed severe limits on the amount of time people under the age of 18 can play them. One way those limits can be enforced is through gameplay data accumulated within games and facial recognition and other biometric data built into the gaming consoles.[13]

11 "You watch your TV. Your TV watches back" by Geoffrey Fowler, *The Washington Post*.
12 "Amazon workers are listening to what you tell Alexa" by Matt Day, Giles Turner, & Natalia Drozdiak, Bloomberg.
13 "The unnerving rise of video games that spy on you" by Ben Egliston, *Wired*.

Things Others Choose

Take a walk through your neighborhood and see how many video surveillance cameras you can readily spy.[14] Don't look too hard, however, or someone will get suspicious. That's because on the other side of many home surveillance cameras is a person viewing you walk by on their smartphone app—and possibly posting your picture on Nextdoor, the neighborhood network app the company claims is used by 1 in 3 U.S. households.

One of the more popular neighborhood surveillance options is the doorbell camera, which is super handy for recording porch pirate package thefts and doorstep conversations no one inside was supposed to hear. They also collect a lot of data about you, your neighbors, and strangers who happen by your front door.

Ring, a company purchased by Amazon in 2018, is the most popular doorbell camera brand. The device records every doorbell press, as well as sounds made within 30 feet and motion occurring in its view. But not just any motion—Ring uses a "human detection" system that can be activated to differentiate human activity through "face and body-shape analysis."[15]

Amazon collects all that data, along with data from the person using the Ring app to review their own footage.[16] Ring also has agreements in place with police to provide doorbell camera video on request—sometimes without notification to the homeowners.[17] That revelation drew criticism from Sen. Ed Markey (D-Mass.), who noted:

> It has become increasingly difficult for the public to move, assemble, and converse in public without being tracked and recorded. We cannot accept this as inevitable in our country. Increasing law enforcement reliance on private surveillance creates a crisis of accountability, and I am particularly concerned that biometric surveillance could become central to the growing web of surveillance systems that Amazon and other powerful tech companies are responsible for.[18]

And you don't have to move beyond your front door for your home activities to be subject to surveillance by others.

Covid-19 forced millions to work from home in early 2020, with many continuing that practice at least part of the time long after. Computers and other equipment brought from work, however, belong to work. Things like email, keyboard keystrokes, web browsing, collaborations with coworkers via team software, video from your webcam, and even the area of your home you're working in are all fair game for monitoring.[19]

A Contemporary Panopticon

In the late 1700s, philosopher Jeremy Bentham conceived of the perfect prison—a space where people were separated in a circular building, with all facing a central tower, the assumption being that all were subject to continual surveillance. The magic of the **panopticon**, as he called it, was that people would assume they were being watched, whether they were or not.

Nearly 200 years later, philosopher Michel Foucault brought the panopticon to contemporary audiences, using it to illustrate the power of technology. He noted, "The Panopticon is a machine for dissociating the see/being seen dyad: in the peripheric ring, one is totally seen, without ever seeing; in the central tower, one sees everything without ever being seen."[20]

---

14 Pun intended.
15 "About the advanced motion detection system used in Ring devices" from Ring Help, General Information.
16 "Amazon's Ring logs every doorbell press and app action" by Leo Kelion, BBC.
17 "Amazon gave Ring videos to police without owners' permission" by Alfred Ng, *Politico*.
18 "Senator Markey's probe into Amazon reveals new privacy problems," press release.
19 "Here are all the ways your boss can monitor you" by Tatum Hunter, *The Washington Post*.
20 Foucault (1975).

This nicely illustrates the contemporary relationship between individuals, technology, and technology companies, particularly the dominance of the Frightful Five. Everything is observed and collected, with no particular goal other than to figure out ways to use and monetize the vast volumes of data.

Foucault, who died before the World Wide Web was invented, came up with a term for this form of data collection: "permanent registration"—data you give freely and can't take back once given. He pointed to its ability to consolidate power, "because it can reduce the number of those who exercise it, while increasing the number of those on whom it is exercised."[21]

Social Psychologist Shoshana Zuboff calls today's parasitic relationship between digital content providers and their customers **surveillance capitalism**, in which data is sucked up wholesale by the machine, fed into an algorithm aimed at predicting future behavior, marketed, and fed back to consumers in the form of hyper-personalized ads. Zuboff said connection is no longer social, inclusive, or aimed at democratizing knowledge. Instead, "digital connection is now a means to others' commercial ends."[22]

Foucault observed that the technology of the panopticon intentionally isolates us from each other. His work particularly summons the nature of social media feeds: "they are like so many cages, so many small theatres, in which each actor is alone, perfectly individualized and constantly visible."[23]

Importantly, the success of the panopticon relies upon our knowledge that we are being continually observed and adjusting our behaviors accordingly. While there's no central control, we make choices every day regarding device use and social media engagement, both of which come with significant costs to privacy. As Zuboff explained,

> This conflict produces a psychic numbing that inures us to the realities of being tracked, parsed, mined, and modified. It disposes us to rationalize the situation in resigned cynicism, create excuses that operate like defense mechanisms ("I have nothing to hide"), or find other ways to stick our heads in the sand, choosing ignorance out of frustration and helplessness.[24]

In the context of the technology-driven panopticon, we're complicit in our imprisonment.

What Does Online Privacy Look Like?

Privacy is in the eyes of the beholder and relative to social context. Well into the 1990s, an American's Social Security Number was a primary form of identification, used for both student identification and driver licensing. Today we think sharing your SSN with anyone other than financial institutions, health care providers, and the government may lead to identity theft.

Privacy on the World Wide Web, smartphones, and apps is complicated by a variety of things, including what we think privacy actually is: individual privacy or aggregate disclosure. The former includes things that relate primarily to you, while the latter relates to finding trends in user-generated content and use data.

Individual privacy relates to wanting limited access to our private content on social media sites, as well as some desire to prevent others from seeing with whom we communicate, what we communicate, and/or where we go on the web.

Individual privacy is the difference between posting something onto someone's social media feed—which is there for all friends and perhaps friends-of-friends to see—versus direct messaging via Facebook or Snapchat, texting on a smartphone, or emailing.

---

21 Ibid.
22 Zuboff (2019), p. 9.
23 Foucault (1975), op. cit.
24 Zuboff (2019), op. cit., p. 11.

Individual privacy also is about being able to gatekeep our online spaces and share varying degrees of ourselves with select people.

Aggregate disclosure relates to what platform service providers do with your personal data. It includes things like Google Trends and rankings of trending words and hashtags on social media sites, where all of the individual pieces of data we share every day are mined in the aggregate[25] for trends.

Individual privacy is violated when someone eavesdrops on a private conversation; aggregate disclosure eliminates that identifiable, individual voice by eavesdropping on all conversations.

Aggregate disclosure is what's happening with personalized advertising. No human is sitting at a computer sieving the web or social media for people who like to run so a running shoe ad can be sent to every site they visit. Instead, running-related terms are algorithmically pooled and the machine sends the ad your way an instant after you've posted your 5K race time.

And how does that running shoe ad end up on every web page you visit, when you only posted the 5K info on Facebook? You're being followed. Like Hansel and Gretel, you're leaving a trail of **breadcrumbs** across the web and social media as you travel. This also extends into the physical world, where the GPS on your smartphone tracks your location. If this trail was like the history setting in your browser, you could just clear it and start a new path. But that's not how web and app tracking works.

The collision between individual privacy and aggregate disclosure comes when a company or the government decides to drill down and isolate the individual from the aggregate. Your web usage patterns somehow flag you as someone deserving of closer scrutiny.[26] And, since the bulk of your online content is hosted by the corporations that may also have created your operating system, browser, and search engine, finding you might not be so very hard to do. It's probably best to always assume you have no online privacy and that every thing you ever post will last in perpetuity. Whatever it is, make it a good one.

Why Don't We Care More about Digital Privacy?

When it comes to privacy versus personalization, we're willing to give up quite a bit of the former in order to have unfettered access to the latter. That risk/benefit trade-off is known as the **Privacy Paradox**.

The term was coined in 2001 by a researcher who found that people would volunteer concerns about digital surveillance, yet were willing to provide personal information if they perceived an advantage to it.[27] The Privacy Paradox is reflected by subsequent studies that demonstrate that, even when given the opportunity to pay a small amount for services that would protect their privacy, people will instead choose free or less expensive services that erode personal privacy.[28]

So, yes, we like free, but there's another reality when it comes to technology and surveillance: We know it's happening, but we love the technology and the idea of reading even a single user agreement before participating is daunting.

Plus, when you're downloading an app or signing up for a service, you've already decided you want to participate. You don't get to negotiate the terms of the user agreement or what parts of your phone an app can access. It's all or nothing.

---

25   Aggregate in this context means taking a mass of individual things and combining them into a whole. Whatever is dominant in the aggregate determines how the mass is perceived. An abundance of pink feldspar in aggregate rock would identify it as pink granite; an abundance of one opinion in aggregate data would identify a trend.

26   The search terms you've recently used on Google, coupled with some vaguely anarchist comments on Facebook, snarky antigovernment tweets, and visits to inflammatory websites, might be enough to get you there.

27   "Studying the internet experience" by Barry Brown, Hewlett Packard Laboratories.

28   The true irony is that people *are* willing to pay a few bucks more for streaming audio and video services so they can skip ads …

A **user agreement** defines the contract between you and the hosting site or app. It also may be called an end-user license agreement or terms of use. The organization's privacy policies generally are a separate, also lengthy document.

While the average length of most user agreements is about 4,200 words, the original iTunes Store user agreement clocked in at nearly 15,000—about one-quarter the size of this book. Meta's combined terms of service offers up a relatively paltry 5,000 words but then suggests you follow links to two dozen or so online documents, including privacy policies and community standards.

Late-night host John Oliver[29] put it best: "If you want to do something evil, put it inside something boring. Apple could put the entire text of *Mein Kampf*[30] inside the iTunes user agreement, and you'd just go 'agree, agree.'"

The Washington Post created a nifty utility that lets you estimate the length of user agreements on a smartphone, and how long it would take you to read them. For example, if you have 72 apps on your phone, the user agreements would span more than 300,000 words and take about 20 hours to read.[31]

Why is privacy such a rare commodity online? It's partly because we are willing to trade our privacy for free services and partly because of the user agreements that exist on every website.

There are two types of user agreements commonly used on the World Wide Web and with apps: click-wrap agreements and browse-wrap agreements.

Click-wrap user agreements are obvious—you're not entering a site until you click on the box that says you agree to the terms of the site. And once you've clicked on the box, it can be really hard to find that terms of use document again, not that you probably read it in the first place. *You want the free service.*

Browse-wrap user agreements are less obvious. If you've been on a site and don't remember having to register or click a box to get in, it's likely a browse-wrap site. Google, for example, is browse-wrap. Its "Privacy & Terms" are linked right from the search home page.

With click-wrap, if you don't explicitly agree, you're not gaining entry to the site. With browse-wrap, you've agreed to a site's terms just by virtue of being on the site. Either way, the site owner is in control of your data at both the individual and aggregate levels.

It's Not the Tech We're Worried About

We use locks to protect our homes, offices, cars, and bikes, passwords to protect our online connections, and biometric locks to protect our phones. We're big on locking things down. What are we protecting these things from? Other people.

Yet we allow limitless surveillance of virtually every aspect of our lives by faceless technology companies. A Pew Research study[32] found that about 8 in 10 Americans agree to a privacy policy at least once a month. And, as we've noted before, it's highly unlikely they're reading them.

Why? Because we're more concerned with perceptions of physical-world risks than virtual ones. Philosophy professor Robert Howell dubbed this the **Peeping Tom effect**, explaining it as "the visceral reaction we have when we think about living, breathing agents observing our private lives, versus how we feel knowing that corporations are collecting the same information and storing it on their computers."[33] In short, we're more worried about a single human looking into our kitchen window than the volumes of personal information continually collected by the smart speaker sitting inside.

29 *Last Week Tonight with John Oliver,* Episode 5: "Net Neutrality." HBO, June 2014.
30 *Mein kampf,* or "my struggle," was the ideological manifesto by Adolf Hitler that was published in the 1920s. The nearly 700 pages of the two-volume series outlined his plans for the rise of the Nazi party in Germany.
31 "I tried to read all my app privacy policies. It was 1 million words" by Geoffrey Fowler, *The Washington Post.*
32 "Americans and privacy: Concerned, confused and feeling lack of control over their personal information" by Brooke Auxier, Lee Rainie, Monica Anderson, Andrew Perrin, Madhu Kumar, & Erica Turner, Pew Research Center.
33 "The Peeping Tom Effect makes us worry about the wrong threats to our privacy" by Robert Howell, OneZero, Medium.

Howell points to this as a psychological bias that draws from the physical threats faced by our ancestors, not the digital ones we face today. He doesn't find that logical or defensible, arguing that we need to adapt sooner than later:

> Only recently have non-conscious cognitive agents entered the picture, so we have no inborn reactions to those sorts of things. That would explain why alarm bells don't sound when we think of violations of our privacy by corporate computers, but it wouldn't vindicate our tendency to ignore them. Psychology aside, we should fear corporations just as much or more, and the longer we suffer from the Peeping Tom effect, the more vulnerable we are to the real threats.[34]

And it isn't just data collection by companies that we ignore. While sitting in a coffee shop with your laptop, you may be conscious of not leaving it or any other valuable alone on the table. Why? Because humans are known to steal things.

But you cheerfully log into the free, public Wi-Fi network. As with many things, this form of digital "free" has a cost in terms to data security.

While public Wi-Fi is a convenience that keeps those without unlimited data plans from paying for data and allows those with laptops to connect to the internet, it's not always secure. Lack of digital encryption increases the possibility that someone else—likely a person in the room with you—can hack into your computer files or observe what's on your screen via the Wi-Fi network.

If all you're doing with Wi-Fi is streaming music or video, you're probably OK. If you're working on anything sensitive, paying bills, banking, engaged in private communication with others, or routinely stay logged into a whole lot of stuff on your device, you need to learn how to use a Virtual Private Network. Or leave device use for when you're on a secure network at home or work.

Then there's that location tracking on your smartphone. Again, the idea of a stranger following you wherever you go is frightening, but we give our devices exactly that access by enabling location services.

For most of us, the benefits of technology use far outweigh privacy concerns. And, while you can't control your privacy, you can take steps to ensure more secure digital engagement. Investigative journalist Julia Angwin suggests assessing your true digital vulnerabilities and doing what you can to protect those. Angwin's threat assessment included the realization that she sets herself up for hacking and identity theft by cutting data security corners:

> Probably my biggest issue is that I have no patience. I often take shortcuts instead of hunkering down to figure out why my technology tools aren't working. As a result, I am liable to leave myself vulnerable to exposure.[35]

It's smart to practice good **digital hygiene**, which means taking advantage of the tools at your fingertips to protect your data—just as you protect the health and safety of your body and belongings. Here are a few tips:

— Don't use public Wi-Fi.

— Turn off location tracking on your phone, except when you really need it.

— Don't rely on services from only one company. For example, you might use Firefox, which is owned by the non-profit Mozilla Corporation, instead of Chrome for your browser, or use a privacy-protecting search engine like DuckDuckGo instead of Google.

---

34   Ibid.
35   J. Angwin (2014). *Dragnet nation: A quest for privacy, security, and freedom in a world of relentless surveillance*, pp. 68–69.

- Use multi-factor authentication for your accounts (it's a pain, but totally worth it).

- Don't use the same password for everything, skip things like your last name and birth year.

- Use a password generator or choose a longer phrase you'll remember, like a song lyric, perhaps replacing letters like E with 3.

- Change your passwords regularly, just as you change the batteries in your smoke detector. (Heck, why not do both at once?)

- Log out of programs and apps you're not using.

- Delete old files.

- Delete apps you don't use.

- Don't respond to messages you're not expecting.

- Don't click on links or open attachments you're not expecting, *regardless of the source.*

- Don't assume an email from a colleague, your boss, or your school is legit, look for credibility cues. Not the correct email domain? Short strange message? Delete.

A Few Classic Case Studies in Online Privacy

Stories of digital privacy breaches by web service providers and the U.S. government are infuriatingly common. They typically are brought to light by privacy advocates who actually read user agreements so you don't have to. The breaches then are circulated by news organizations and spread via social channels. Most deal with issues relating to aggregate disclosure.

One of the earliest and most notable is Facebook's **emotional contagion study,** which was revealed in summer 2014 after a team of Facebook and academic researchers published a journal article about it. The manipulation itself took place in 2012 and involved more than half a million Facebook users. The study setup was that Facebook algorithmically filtered out everything but positive/happy posts in one set of user feeds and filtered out all but more negative/depressing posts in the other. The researchers wanted to see if manipulated users would "catch" the emotional vibe presented in their feeds and respond by making positive or negative posts.[36]

The outcome isn't as significant as the backlash when news of the study broke. Privacy advocates used this example to say, "Hey, this is why user agreements matter." Then Facebook itself announced that the option of using user data for experimental psychological research was *not* in its user agreement at the time of the study—the company added it four months after the experiment ended. There also was a brief flurry of conspiracy theories that connected the study to the U.S. military's interest in the generation of civil unrest.

Whatever the purpose, Facebook effectively duped hundreds of thousands of its users with no apparent consequence to the company.

---

36 "Experimental evidence of massive-scale emotional contagion through social networks" by Adam Kramer, Jamie Guillory, and Jeffrey Hancock, National Academy of Sciences.

Whether or not the experiment had consequences to individual Facebook users depended on whether a person's feed was included, which randomly selected emotional manipulation was applied to the feed, and an individual's state of mind. In other words, if you were in the downer group and already having a bad day, Facebook probably made it far worse.

A decade wiser, it's not hard to imagine such manipulation as being dangerous. Yet there were virtually no consequences to Facebook, which gave the company no incentive to stop "innovating" or experimenting on its customers.

While American web users may be relentlessly forgiving, American platform service providers have encountered serious pushback in Europe for privacy breaches.

A **"Right to be Forgotten"** ruling in May 2014 from the European Union's Court of Justice has been particularly vexing for search giant Google. The ruling allows European citizens of EU member countries to petition Google to prevent information about them from being indexed by its searches. Perhaps most difficult for Google is the requirement that the company review requests on a case-by-case basis to determine if the information is "inaccurate, inadequate, irrelevant or excessive."[37] This requires the algorithmically driven search engine to rely on the judgment of expensive human filters.

Legacy media organizations have expressed great frustration with Google's response to the "Right to Be Forgotten" ruling, which has resulted in a large number of news articles being blocked from search results. Other concerns have been expressed about the ability of public relations firms to request removal of information about corporate clients.

Europe also has reined in **Google Street View cars** for their privacy-invading potential far more aggressively than the United States. The street view function of Google Maps is disabled or limited in many countries for various reasons.

For example, Google's Street View cars tapped into unsecured home and public wireless networks in about 30 countries, gathering user data. Google admitted that the practice went on from 2008 to 2010 but was not intentional.

In the United States, the data tapping resulted in Google being fined, ordered to destroy the data, and end the practice. The company appealed that ruling all the way to the Supreme Court, which in 2014 agreed with lower courts that Google had violated the Wiretap Act.[38]

While stories of *individual* privacy breaches don't always make national or international news—especially if they don't involve a celebrity—they do happen.

A memorable one involved a young woman who lived in Washington, D.C., and blogged about her sex life *for two weeks* in 2004. Jessica Cutler regaled her friends with *Sex and the City*-type commentary about her relationships with six men, scattering details about them throughout ongoing blog entries. She also dropped enough bits of information across her blog entries about herself that she was identifiable.

When her blog was linked to by a very popular Washington, D.C., blogger, Cutler stopped posting and deleted it. It was too late, however, and she was fired for inappropriate use of a computer that belonged to her employer, a U.S. senator.

The blog's trail of details about one sex partner, who apparently enjoyed spanking, also meant that *he* was identifiable. He sued her for publication of private facts and invasion of privacy, seeking $20 million in damages. The civil lawsuit was dismissed five years later, with the ex-lover ordered to pay $14,000 in legal fees to Cutler's attorney.[39]

---

37   The full text of the ruling is available online by searching for C-131/12 on the court's archive site: curia.europa.eu.

38   The not-for-profit Electronic Privacy Information Center (epic.org) is an excellent source of more information about electronic privacy.

39   The blogger, Jessica Cutler, fictionalized the experience in a novel called *The Washingtonienne*.

There's a cautionary tale here: Don't use work devices for personal stuff and, if you're going to publicly share private things about other people—particularly on social media—don't. Just don't.

For an individual, a breach of personal privacy can be a life-changing mistake. For a Platform Service Provider, both privacy breaches and aggregate disclosure are business as usual. As tech journalist Julia Angwin noted,

> This is the tragic flaw of "privacy" in the digital age. Privacy is often defined as freedom from unauthorized intrusion. But many of the things that *feel* like privacy violations are "authorized" in some fine print somewhere.[40]

Information is power. Anyone who holds a vast amount of information about us has power over us.[41]

Who Else Is Interested in What You're Doing Online?

While Platform and Internet Service Providers are not particularly interested in you knowing how they use your data, they are interested in you knowing who else may be asking for it.

Service providers are asked for a variety of information every day from both government officials and private companies. The requests range from user and account information, to all content created, to access to user devices, to all of the above.

Transparency reports reflect that the companies have three choices when they receive these requests. They can:

– Accept the request and provide all information available,

– Accept the request and provide some information, or

– Reject the request and provide no information.

Most tech companies—from social media to phone service providers—provide annual transparency reports that offer widely varying degrees of information. Transparency reports may be found by typing "Transparency report for …" into a search engine, replacing the " … " with the desired service brand.

There's no standardization to the reports, so comparing them is a bit like comparing eggplants and mustard seeds—too much or too little data to evaluate. One interesting trend does emerge, however, in reviewing transparency data from the aforementioned companies: The U.S. government requests far more access to account, user, and device information than *any other country*.

A 2022 report from the American Civil Liberties Union found that the U.S. Department of Homeland Security secured mobile app location data from hundreds of millions of phones, paying about $20,000 per year for that access.[42]

And, in the last half of 2021, Facebook received nearly 400,000 requests for user information from across the globe, with roughly 30% coming from the United States.[43] To put that in perspective, U.S. Facebook accounts make up only 6% of the social media company's global use. And Facebook also said "yes" to U.S. requests 88% of the time, compared to the global average of 73%.

---

40   Angwin (2014), op. cit. p. 3.
41   Ibid., p. 19.
42   "Homeland Security records show 'shocking' use of phone data, ACLU says" by Alfred Ng, Politico.
43   Updated data can be found in the "government data requests" section of Meta's Transparency Center.

The government has varied reasons for interest in people. Sometimes it's what you're doing, sometimes it's who you associate with—and social media is an ideal way to find those things out. We provide a lot of information and evidence about both via what we post, share, and comment on across social media, what we search for on the web, and where our devices say we go.

But technology doesn't place the power only in the hands of those who gather data about us and share it with advertisers and the government.

The flipside to ubiquitous surveillance is equally ubiquitous *sousveillance*.[44] While surveillance comes from above—government, police, and other formal institutions—sousveillance comes from below. It's best illustrated by the use of smartphones to record video of governmental agents doing wrong.

Sousveillance is what 17-year-old Darnella Frazier was doing when she filmed George Floyd's death at the hands of Minneapolis police officers in 2021. That viral video resulted in global attention to police brutality, as well as the suspension, prosecution, and conviction of all officers involved. Frazier's quick thinking and courage also earned her a special award from Pulitzer, the highest journalism award in the U.S., which recognized the crucial role individuals can play in bringing attention to injustice.

The truly remarkable thing about digital technology is that it's also a source for accountability—we can freely call out our public officials and media, as well as widely share evidence of wrongdoing. Today's tech doesn't place the power only in the hands of those who gather data about us and share it with advertisers and the government, it also gives us the power to make a difference.

Summary

Perceptions of what privacy is and what kinds of things should be kept private have evolved greatly since the advent of social media and the invention of smartphones. Most Platform Service Providers allow individuals to regulate what others see in our social media feeds. This allows us to share parts of our lives with others, while sometimes limiting who can view particular information we post online. In turn for free space on which to post our content, Platform Service Providers collect astronomical quantities of personal data. Some data are sold to advertisers, who then target us with advertising that relates to our personal lives, professional interests, and recreational pursuits. Individual data also are gathered by Platform Service Providers, pooled in astronomical quantities, and mined for trends. These aggregate data help us—and advertisers—figure out what is deemed culturally relevant at any given time. The amount of personal information available to Platform Service Providers also has led to significant privacy breaches, such as studies that manipulate user content and then gauge reaction.

Key concepts

Aggregate disclosure
Breadcrumbs
Browser fingerprint
Browse-wrap
Click-wrap
Cookies
Digital hygiene
Digital surveillance
Emotional contagion study
Individual privacy

---

44 This term was coined by inventor Steve Mann and popularized by sci-fi author and internet security expert David Brin with his 1998 book *The transparent society: Will technology force us to choose between privacy and freedom?*

Internet of Things
Internet Protocol address
Native apps
Panopticon
Peeping Tom effect
Privacy Paradox
Right to be Forgotten
Sousveillance
Surveillance capitalism
Trackers
Transparency reports
User agreements

References

Angwin, J. (2014). *Dragnet nation: A quest for privacy, security, and freedom in a world of relentless surveillance*. New York, NY: Times Books.

Auxier, B., Rainie, L., Anderson, M., Perrin, A., Kumar, M., & Turner, E. (2019, November 15). Americans and privacy: Concerned, confused and feeling lack of control over their personal information. *Pew Research Center*. Retrieved from https://www.pewresearch.org/internet/2019/11/15/americans-and-privacy-concerned-confused-and-feeling-lack-of-control-over-their-personal-information/

Brown, B. (2001, March 26). Studying the internet experience. *Hewlett Packard Laboratories*. Retrieved from https://www.hpl.hp.com/techreports/2001/HPL-2001-49.pdf

Burgess, M. (2022, February 26). The quiet way advertisers are tracking your browsing. *Wired*. Retrieved from https://www.wired.com/story/browser-fingerprinting-tracking-explained/

Cutler, J. (2006). *The Washingtonienne*. New York, NY: Hyperion Books.

Day, M. Turner, G., & Drozdiak, N. (2019, April 10). Amazon workers are listening to what you tell Alexa. *Bloomberg*. Retrieved from https://www.bloomberg.com/news/articles/2019-04-10/is-anyone-listening-to-you-on-alexa-a-global-team-reviews-audio#xj4y7vzkg

Egliston, B. (2022, February 1). The unnerving rise of video games that spy on you. *Wired*. Retrieved from https://www.wired.com/story/video-games-data-privacy-artificial-intelligence/

Foucault, M. (2008). Panopticism (A. Sheridan, Trans.). In *Discipline and punish: The birth of a prison*. New York, NY: Pantheon. Reprinted in *Race/ethnicity: Multidisciplinary global contexts*, pp. 1–12 (2008). Indiana University Press. Retrieved from https://muse.jhu.edu/article/252435/pdf

Fowler, G. A. (2019, May 28). It's the middle of the night. Do you know who your iPhone is talking to? *The Washington Post*. Retrieved from https://www.washingtonpost.com/technology/2019/05/28/its-middle-night-do-you-know-who-your-iphone-is-talking/

Fowler, G. A. (2019, September 19). You watch your TV. Your TV watches back. *The Washington Post*. Retrieved from https://www.washingtonpost.com/technology/2019/09/18/you-watch-tv-your-tv-watches-back/

Fowler, G. A. (2022, May 31). I tried to read all my app privacy policies. It was 1 million words. *The Washington Post*. Retrieved from https://www.washingtonpost.com/technology/2022/05/31/abolish-privacy-policies/

Howell, R. (2019, November 7). The peeping tom effect makes us worry about the wrong threats to our privacy. *OneZero, Medium*. Retrieved from https://onezero.medium.com/the-peeping-tom-effect-646f1c60fb4f

Hunter, T. (2021, September 24). Here are all the ways your boss can monitor you. *The Washington Post*. Retrieved from https://www.washingtonpost.com/technology/2021/08/20/work-from-home-computer-monitoring/

Hunter, T. & Kelly, H. (2022, June 24). With Roe overturned, period-tracking apps raise new worries. *The Washington Post*. Retrieved from https://www.washingtonpost.com/technology/2022/05/07/period-tracking-privacy/

Kelion, L (2020, March 4). Amazon's Ring logs every doorbell press and app action. *BBC*. Retrieved from https://www.bbc.com/news/technology-51709247

Kramer, A. D. I., Guillory, J. E., & Hancock, J. T. (2014, March 25). Experimental evidence of massive-scale emotional contagion through social networks. *Proceedings of the National Academy of Sciences*. Retrieved from http://www.pnas.org/content/111/24/8788.full

Ng, A. (2022, July 13). Amazon gave Ring videos to police without owners' permission. *Politico*. Retrieved from https://www.politico.com/news/2022/07/13/amazon-gave-ring-videos-to-police-without-owners-permission-00045513

Ng, A. (2022, July 18). Homeland Security records show "shocking" use of phone data, ACLU says. *Politico*. Retrieved from https://www.politico.com/news/2022/07/18/dhs-location-data-aclu-00046208

Oliver, J. (2014, June). Net neutrality [Segment of television series episode, 10.00 to 10:31]. *Last Week Tonight with John Oliver*. HBO. Retrieved from https://www.youtube.com/watch?v=fpbOEoRrHyU#t=601

Saad, L. (2022, June 20). Americans have close, but wary bond with their smartphone. Gallup. Retrieved from https://news.gallup.com/poll/393785/americans-close-wary-bond-smartphone.aspx

Senator Markey's probe into Amazon reveals new privacy problems. (2022, July 13). Press release, Ed Markey U.S. Senator for Massachusetts. Retrieved from https://www.markey.senate.gov/news/press-releases/senator-markeys-probe-into-amazon-ring-reveals-new-privacy-problems

Song, V. (2021, December 21). Amazon Halo review: The Fitbit clone no one asked for. *The Verge*. Retrieved from https://www.theverge.com/22834452/amazon-halo-view-review-fitness-trackers

Zuboff, Shoshana (2019). *The Age of Surveillance Capitalism: The Fight for a Human Future at the New Frontier of Power*. New York, NY: PublicAffairs.

CHAPTER 16

Can't Put the Genie Back in the Bottle: Now What?

Figure 16.1. An optimistic view—looking forward in the same direction, without mediation. Pearls Before Swine © 2016 Stephan Pastis. Reprinted by permission of Andrews McMeel Syndication. All rights reserved.

In 2005, two journalists produced a nine-minute film that focused on the rising tension between the democratic imperative of a free press and the rise of algorithm-driven entertainment.

EPIC 2015[1] was supposed to be a wake-up call to journalists and news organizations that journalism was in danger—a stark warning of what the future might look like within a decade if the rising power of digital media companies was not challenged.

Of specific concern was **news aggregation**, the free dissemination of journalist-produced content across social media and search engines, and the impact it would have on the media companies that paid journalists to do that work.

Spoiler alert: the film producers were right.

EPIC 2015 starts its timeline with 1989 and the invention of the World Wide Web, progressing through the launch of Amazon (1994), Google (1998), Google News (1999), and one of the earliest social media platforms, Friendster (2002).

In response to Google News, in 2003 Microsoft launched Newsbot, an automated news service that allowed users to choose topics of interest and have stories selected from more than 4,000 newspaper websites. At this point, Microsoft's Hotmail was one of the few email services that were free to customers, so Google dropped its own competitor, Gmail, into the marketplace in 2004.

This is where EPIC 2015, guessing the future a decade ahead, begins to deviate from reality—*with eerie precision*. Here are some key predictions, along with the reality:

- 2005—*Microsoft buys Friendster.*
  Reality check: That didn't happen, but Facebook, which was just entering the market, ultimately did buy—and kill—Friendster in 2010.

- 2005—*Apple releases the wifiPod*, "a portable media player with integrated camera, that can send and receive podcasts and images on the go."[2]
  Reality check: Apple released the both the iPod Touch and iPhone in 2007. The above prediction is closer to the former.[3]

- 2005—*Google creates the Google Grid*, "a universal platform offering an unlimited amount of space and bandwidth that can be used to store anything."
  Reality check: This essentially describes Google Workspace[4] and Apple's iWork,[5] which are direct competition to the decades-dominant Microsoft Office.[6]

- 2007—*Microsoft creates Newsbotster*, a social news network built on Friendster's technology.
  Reality check: This envisioned social news network incorporates key features of Reddit, Twitter, and Facebook, including allowing users to comment and share.

- 2008—*Google and Amazon merge to form Googlezon*, coupling the Google Grid with Amazon's personalized recommendations.
  Reality check: Didn't happen. Didn't need to. Both companies were already dominating different areas of the web.

---

1   EPIC 2015 was produced in 2005 by Robin Sloan and Matt Thompson and is available on YouTube (https://youtu.be/OQDBhg60UNI).
2   Ibid.
3   Seriously, who but Steve Jobs could have foreseen the cultural disruption sparked by the iPhone?
4   Google Workspace = Gmail, Calendar, Drive (cloud storage), Meet, Docs, Sheets, Slides, Chat, and Sites.
5   iWork = iCloud (cloud storage), Pages, Numbers, and Keynote.
6   Microsoft Office = Outlook, OneDrive (cloud storage), Word, Excel, PowerPoint, OneNote, and Teams.

- 2010—***The news wars begin,*** "notable in that they involve no actual news organizations"— the battle is between Microsoft and Googlezon. EPIC 2015 foretells Googlezon coming out on top, with an algorithm that strips sentences and facts from all news content online and recombines them into stories customized for each user.

  Reality check: The prediction of news wars without news media is exceedingly accurate, but today is more of a clash of the titans than EPIC 2015 predicted.

  The unofficial war on news organizations, which is perhaps a bit more accurate, started in about 2009, when Google Fast Flip was launched to cinch Google's control over news aggregation. Fast Flip was short lived, as it stripped native ads from news organizations and replaced them with ones generated by Google's own ad service, DoubleClick. Not only did the user not see the ads that helped pay the news organization's bills, but they also didn't have to click through to the home site.

  The "news wars" began in earnest in 2015 when Apple launched the one-two punch of iPhone Operating System 9 and Apple News. The pair featured robust ad blockers designed to hurt Google's market-dominant DoubleClick ad server. As an aptly-titled article[7] in *The Verge* noted,

  > …there's no other company that's managed to monetize the web quite like Google … users search for things using Google, see Google search ads, and then land on content that is further monetized by Google DFP and its ad exchange. This is basically the foundation of Google's entire business: Google makes the lion's share of its money on search, and Google search doesn't work if the web isn't searchable, so Google has a huge interest in making the web profitable for media companies, so they can search all that content.

  Apple's effort worked only for iPhone users, who didn't have an option at that point to use anything but the iPhone's native web browser, Safari. With Chrome dominant on desktops and supporting ad blocking, and more eyeballs moving to mobile, this was "Apple's attempt to fully drive the knife into Google's revenue platform."[8]

  Facebook also had a role in this battle, incentivizing news sharing on its app by elevating reputable news content in the feed. This went south pretty quickly, with Facebook's newsfeed algorithm being manipulated in the run up to the 2016 U.S. presidential election, as noted in Chapter 8.

  The mid-decade news wars resulted in news organizations sharing little of the digital advertising bounty, with Apple, Facebook, and Google competing for the spoils. Each had a particular area of dominance: Apple, the iPhone, Facebook, its app, and Google, the web.

- 2011—***The "slumbering Fourth Estate awakens"*** and sues Googlezon, claiming the company's "fact-stripping robots are a violation of copyright law." It's too late. The case, brought by legacy media like *The New York Times*, makes it to the Supreme Court, which rules in favor of Googlezon.

  Reality check: Despite clear warnings from a variety of sources, including congressional hearings focused on the impacts of Big Tech, this hasn't happened. It may well be too late.

- 2014—***Googlezon unleashes EPIC***, the Evolving Personalized Information Construct. EPIC produces an algorithm-driven custom content package for each user, focusing on individual demographics, choices, consumption habits, interests, and social connections. The content

---

7   "Welcome to hell: Apple vs. Google vs. Facebook and the slow death of the web," by Nilay Patel.
8   Also from EPIC 2015, which you, too, can find on YouTube (https://youtu.be/OQDBhg60UNI).

largely comprises individual reports, including images and video. All users can contribute and some get paid a cut of Googlezon's ad revenue based on the popularity of their contributions.

> At its best, "EPIC is a summary of the world—deeper, broader, and more nuanced than anything ever available before. But at its worst, and for too many, EPIC is merely a collection of trivia, much of it untrue, all of it narrow, shallow, and sensational."[9]

Reality check: YouTube and TikTok each pay their top creators, depending on the popularity of posts, and that's a pretty apt description of a lot of social media.

Just a reminder that the journalists who produced EPIC 2015 wrote the script in 2004. Overall, they engaged in spectacularly accurate guessing. This prophetic little video was produced right around the time that YouTube was launching in 2005. Facebook at that point was just getting started and Twitter (2006), the iPhone (2007), and Instagram (2010) were not yet commercially available.

EPIC 2015 is available on YouTube, where it has a surprisingly small number of views. At fewer than 150,000, its views are eclipsed by a Vine clip of a micro pig running through the grass—actually posted in 2015—which has 325,000 views and counting. That actually says a lot about where we are today.

In 2014, a development team in West Germany updated EPIC 2015, using artificial intelligence to cobble together video clips better reflecting the decade of tech evolution. It ends with one more unheeded warning: "Perhaps there was another way."[10]

Finding Our Way to Personal Responsibility

We are well on our way to a major evolution in how America works. The future is a moving target and there's no telling what great wonder is ahead. The only thing that is certain is more technology and more change. And, in all likelihood, the past will escort us right into the future, as it always has, and we will continue to accumulate media platforms and the new social problems they pose.

Choosing how to engage and with whom will be more important than ever. So will paying attention to how others engage and, importantly, who's *not* engaging.

The notion of digital literacy is compelling because of the power it offers. All of us make choices about how much we need to understand the technologies that we use. How many people understand how a car does what it does to get us from one place to another?

But in the digital sphere there is much more at stake than might appear. If we make an investment in our growing digital literacy, we come to a better understanding of our relationship to all of what we do—and who is engaging us—when we adopt and use these technologies.

One take on this was offered by media theorist Marshall McLuhan, who noted "there is absolutely no inevitability as long as there is a willingness to contemplate what is happening."[11]

Hopefully reading and thinking about the complexities of this tech has given you a glimpse of what you can do to gain more power over your experiences, to think about what happens when you go online. You go to a site, watch a video; when you get to the end of that video, the website offers you another option. No one tells you that this suggestion is the result of all of the information you have unknowingly given the site so it can keep you there.

9   Ibid.
10   Museum of Media History, Berlin. (https://museumofmediahistory.com/)
11   McLuhan and Fiore (1967), p. 25.

It makes an enormous amount of difference in your experience if you understand what that website wants to do with you. Does it want to make money? Does it want to collect and then sell data about your choices? Does it want to convince you that paying a subscription or an access fee will be worth it?

After all it's your money and it's your time. Sadly, for most of us there are limited amounts of each of these commodities, yet both play a big role in our desire for happiness, fulfillment, and a better quality of life.

Digital (That) Divides (Us)

Digital engagement today allows some people to approach the edge of humanity—living beyond the limitations of physical existence—while others remain firmly tethered to the physical world.

Sci-fi author William Gibson pointed to tremendous global wealth inequities and uneven distribution of technology as having the potential to sow epic division as it fosters a post-human state:

> Consider the health options available to a millionaire in Beverly Hills as opposed to a man starving in the streets in Bangladesh. The man in Beverly Hills can, in effect, buy himself a new set of organs. I mean, when you look at that sort of gap, the man in Bangladesh is still human. He's a human being from an agricultural planet. The man in Beverly Hills is something else. He may still be human, but he, in some way, I think he is also posthuman. The future has already happened.[12]

Think about that last thought for a moment: the future has already happened. That makes sense if you think about the difference between a closed video game and an open one that allows you to explore the virtual world as you choose. You have boundaries to that exploration either way, but closed games require you to move through the story in a linear fashion, while open games have the freedom of being non-linear.

A closed game allows you to make choices, complete tasks, and proceed according to them, but the ending is preconfigured by the game developer. *The future has already happened.*

In an open game, there are possibilities that have not yet been anticipated. In that case, the future remains … well, in the future.

On one hand someone else has already made critical decisions for you; on the other, you have some autonomy to decide where you want go. Either way, you have a choice about which type of virtual world you immerse yourself in.

But sometimes we don't really consider what's happening, because we're viewing it through the lens of what *was*. McLuhan put it like this:

> When faced with a totally new situation, we tend always to attach ourselves to the objects, to the flavor of the most recent past. We look at the present through a rear-view mirror. We march backwards into the future.[13]

Consideration of the present and future require us to be more digitally literate. This is something we should want for ourselves *and* share with people we care about. Digital literacy encourages us to break free from "our deeply embedded habit of regarding all phenomena from a fixed point of view,"[14] think about what we're seeing, what others are seeing, and how it affects us all.

---

12   Gibson, author of the influential 1984 sci-fi book *Neuromancer*, which heavily influenced the 1999 film "The Matrix," made this observation in "Cyberpunk: The Documentary." The 1990 film was directed by Marianne Trench and also features an interview with psychedelic drug guru Timothy Leary. It's available on YouTube.

13   McLuhan & Fiore, op. cit., pp. 74–75.

14   Ibid., p. 68.

Why Aren't We Paying More Attention?

When you get in your car and drive, do you think about the effects that has on the environment? When you eat fast food, do you think about the effects it may have on your body?

We can bring similar questions to the pursuit of digital literacy, to conceptualize the larger impact of the media choices we make. Media researcher Antonio López has spent a career helping us think about this question, dubbing it "colonization of our attention." He said we need to reframe "media" as ecomedia, to offset our tendency not to consider thoughts and ideas as having physical weight in the world. This allows digital media, led by Big Tech—to exert influence without sustained scrutiny.

> Renaming media ecomedia addresses media's ecological opacity (in the sense of unseen, unrecognized, ephemeral, hiding in plain sight, and taken for granted). Ecomedia reframes media as ecological media; that is, media are a material reality that are in, and a part of, our environment in the broadest sense(s). There are no media that are inseparable from their material conditions in the environment that produce them.[15]

This is a powerful suggestion—that we secure the actual material relationship from media into the world. But that is only one of several issues that we should pay more attention to when it comes to our media environment.

That also means critiquing how social media becomes a vector for misinformation about climate change, democratic election, and public health. Your digital literacy gives you the tools and the power to know the difference between science and misinformation, observable facts from propagandist fiction.

Recent scholarship has suggested that our lack of understanding of the social effects of new technology is a danger to both science and democracy.[16] A team of 17 researchers from a range of disciplines—philosophy to biology—urged a crisis-discipline approach that reaches across fields to address tech platforms as an urgent concern, rather than passively watching as they are overrun by fast-spreading misinformation and disinformation.

And yet, there's still a popular belief that these problems have always existed, but it's the technology that makes them more visible. Infectious disease expert Carl Bergstrom, one of the researchers calling for a more rapid response to technological transformation, responded to the technology-as-a-mirror argument like this:

> This should be a familiar argument. This is what Big Tobacco used, right? They said, "Well, you know, yeah, sure, lung cancer rates are going up, especially among smokers—but there's no proof it's been caused by that."

> And now we're hearing the same thing about misinformation: "Yeah, sure, there's a lot of misinformation online, but it doesn't change anyone's behavior." But then all of a sudden you got a guy in a loincloth with buffalo horns running around the Capitol building. [17]

At the same time, the problems inside tech companies are becoming more widely understood. Whistleblowers like Francis Haugen from Facebook are pointing out how such companies specifically target younger customers. For example, Haugen testified before Congress that Instagram makes issues around body image worse—as a design feature, not a bug or unexpected outcome. Companies like Meta, according to the former employee's testimony, know more about the negative effects of their sites than they admit.

---

15   Lopez (2021), p. 9.
16   Bak-Coleman, Alfano, Barfuss, and Weber (2021).
17   "Why some biologists and ecologists think social media is a risk to humanity" by Shirin Ghaffary, S., *Vox Recode*.

Regulation, Anyone?

This testimony leads to one of the points of possible agreement across political divides—that Big Tech needs to be reined in, perhaps even broken up into smaller companies (re-read Chapter 3, if you need a refresher).

So, are there solutions in regulation? Some media industries, like music, gaming, and theatrical films, have developed forms of self-regulation, as with ratings systems. These industries would much prefer to do this as an industry, rather than have regulations imposed from a governmental entity.

But regulation has not been a popular option for the most part. This could be because of an ideological position that the competitive, capitalist marketplace will solve such problems. It could also be because of the power of lobbyists, who represent a massive corporate investment in beating back regulation. There is a resulting lack of momentum for regulatory approaches to be a significant part of the solution.

Self-regulation, then, may be the only workable option—we have to do it ourselves.

Everything in this book is geared toward helping you become better informed about digital literacy issues, which means you also have become better informed about how powerful these experiences are. If a media company's goal is to keep you paying attention, we need to be aware of how to defend ourselves from their efforts.

So one's growing digital literacy is actually a great tool for overcoming this power. Sometimes that work is hard, and sometimes it's easy. Sometimes we have to see ourselves seeing what we are doing—and assert control, to say "stop going down this rabbit hole." Maybe even out loud.

Where Are We Now?

Technology can be very good. It allows us greater access to information than ever before and the ability to build our own communities of spirit to augment our communities of space.

Technology can be not so good. There will always be people who technology does not benefit and problems it will not solve.

Among all the rapid change that has brought us to a digital society—which is both a utopia and a dystopia—is the possibility of *getting it,* of understanding how it works.

The significance of digital literacy is the power that comes with seeing technology for what it is and seeing ourselves for what we are when we are within it. That supports a belief in the power of literacies of all kinds as a motivation we can share. And the sharing that takes place in the digital world helps make the world a better place because of what you can help to spread.

Even if this book just hinted at how one person—you—can come to understand how to work the digital world instead of just being worked by it, then you are on your way.

For tech companies, the power is in getting you to trust that using the digital tools is enough; that you should never open them up to see how they work.

So we ask, for whom is that knowledge a threat? It's a threat to the people and information technology companies that want to keep you out of the game, who want our digital experience to only happen on their terms.

When we educate ourselves—take an information literate perspective—and ask questions about who gets to decide and why, we take back some of the power. That's striking the critical balance between Panglossian cyber-enthusiasm and Pandorian cyber-pessimism[18] by making the technology work for rather than against you.

Not only that, but you can encourage the same things in others. That's the secret of knowledge: You don't want to make it scarcer; it's a power you want to share, to spread, to see in everyone.

18   "Which technology and which democracy?" by Benjamin Barber, Democracy and Digital Media Conference, MIT.

You may want to see it in institutions, like schools, starting as early as people start using the web and smartphones (which is really, really early). You might want to engage in a discussion about digital information rights on the web. You might gain an audience and find yourself in the company of like-minded people who want to spread the idea even further.

By "like-minded," we don't mean people who are *just like you*. Part of why digital literacy is so powerful is because the others you end up engaging with may not be like you—and yet you want them to succeed the same way you do. That's the power that comes when we are able to use these complex digital tools to ask questions and to see the value in the questions of others.

And Then What?

While it's unlikely we're on the cusp of a neo-Luddite revolution, there is increasing awareness that the more we live through our screens, the more isolated we become.

Richard Williams, a rapper better known as Prince Ea, addressed that concern in a 2015 video on his YouTube channel, titled "Can We Auto-Correct Humanity?"[19] It quickly went viral and, today, has more than 23 million views. Here's an excerpt:

> Kinda ironic, ain't it? How these touch screens can make us lose touch.
>
> You do have a choice, yes. Take control or be controlled: make a decision.
>
> No longer do I want to spoil a precious moment by recording it with a phone—I'm just gonna keep them.
>
> I don't wanna take a picture of all my meals anymore—I'm just gonna eat them.
>
> I'm so tired of performing in the pageantry of vanity, and conforming to this accepted form of digital insanity.
>
> Call me crazy, but I imagine a world where we smile when we have low batteries—'cause that will mean we'll be one bar closer … to humanity

The liner notes beneath explain the video's underlying message. Prince Ea noted that his intent is not for people to desert social media platforms or destroy their smartphones, but to "be balanced, be mindful, be present, be here."

And Hugh McGuire, posting on Medium, tells the story of going to his 2-year-old daughter's dance recital and being just one of many parents involved in digital witnessing—taking in the event through a screen and documenting it, rather than just enjoying it.

As he watched, he felt the itch to check email and Twitter on his smartphone—just in case. From that experience, McGuire made the following observation:

> It makes me feel vaguely dirty, reading my phone with my daughter doing something wonderful right next to me, like I'm sneaking a cigarette.
>
> Or a crack pipe.
>
> One time I was reading on my phone while my older daughter, the four-year-old, was trying to talk to me. I didn't quite hear what she had said, and in any case, I was reading an article about North Korea. She grabbed my face in her two hands, pulled me towards her. "Look at me," she said, "when I'm talking to you."
>
> She is right. I should.[20]

---

19   Search for it online and watch it until the end. You can do it. It's 3 minutes and 27 seconds of your life (and it has more than 23 million views, so he must have done something right). Hang around and check out some of his other work.
20   Hugh McGuire's piece for Medium is entitled, "Why Can't We Read Anymore?" and was published April 22, 2015. Search for it online. Read it until the end … if you can.

And artist Stephan Pastis, who draws the comic "Pearls Before Swine" featured throughout this book, thinks the relentless digital cacophony destroys creativity. He wonders how people—young and old—can find inspiration with so much distraction.

> When I was a kid I had seven (TV) channels. I had Mad Magazine. I had some Peanuts books … I had no cable. I had nothing beyond those channels. And the result was I sat in my room and I wrote or I drew. So if I'm that kid now, as I still am in some ways, I go on YouTube and fall down the rabbit hole and watch everything. That kid has a thousand channels. He has a limitless option on the internet. He has video games. That is a hard kid to reach … You need quiet time or you can't hear anything.[21]

And media theorist Douglas Rushkoff, who pointed toward technology as hurtling humans toward an abyss, largely of our own making:

> Faced with a networked future that seems to favor the distracted over the focused, the automatic over the considered, and the contrary over the compassionate, it's time to press the pause button and ask what all this means to the future of our work, our lives, and even our species.[22]

Williams, McGuire, Pastis, and Rushkoff are part of the growing chorus reminding us of the strengths and weaknesses, benefits and drawbacks, of digital information. And all are calling for moderation.

There's nothing inherently wrong with your engagement with technology, as long as you have a degree of digital literacy that helps you think about why you use digital media as you do.

As Marshall McLuhan noted, "now we have to adjust, not to invent. We have to find the environments in which it will be possible to live with our new inventions."[23]

So turn to digital media and technology for the boundless knowledge and efficiency they bring to your life. Don't turn to technology to stave off loneliness or boredom.

Enjoy the wonder that is digital. Then turn it off, look up, and choose to live in the moment.

References

Bak-Coleman, J. B., Alfano, M., Barfuss, W., & Weber, E. U. (2021). Stewardship of global collective behavior. *The Proceedings of the National Academy of Sciences, 118*(27). Retrieved from: https://www.pnas.org/doi/10.1073/pnas.2025764118

Barber, B. R. (1998, May). Which technology and which democracy? *Democracy and Digital Media Conference, MIT.* Retrieved from http://web.mit.edu/m-i-t/articles/barber.html

Ghaffary, S. (2021, June 26). Why some biologists and ecologists think social media is a risk to humanity. *Vox Recode.* Retrieved from: https://www.vox.com/recode/2021/6/26/22550981/carl-bergstrom-joe-bak-coleman-biologists-ecologists-social-media-risk-humanity-research-academics

Lopez, A. (2021). *Ecomedia literacy; Integrating ecology into media education.* New York, NY: Routledge.

McGuire, H. (2015, April 22). Why can't we read anymore? *Medium.* Retrieved from https://medium.com/@hughmcguire/why-can-t-we-read-anymore-503c38c131fe

McLuhan, M. & Fiore, Q. (1967). *The medium is the massage: An inventory of effects.* New York, NY: Bantam Books.

Patel, N. (2015, September 17). Welcome to hell: Apple vs. Google vs. Facebook and the slow death of the web. *The Verge.* Retrieved from: https://www.theverge.com/2015/9/17/9338963/welcome-to-hell-apple-vs-google-vs-facebook-and-the-slow-death-of-the-web

Rushkoff, D. (2011). *Program or be programmed: Ten commandments for a digital age.* New York, NY: Soft Skull Press.

Sloan, R. & Thompson, M. (2005). *EPIC 2015.* Retrieved from: https://youtu.be/OQDBhg60UNI

Transcript from a video interview with Stephan Pastis. (2016). *Reading Rockets.* Retrieved from https://www.readingrockets.org/books/interviews/pastis/transcript

Williams, R., as Prince Ea. (2015). *Can we autocorrect humanity?* [Video]. Retrieved from https://youtu.be/dRl8EIhrQjQ

21 "Transcript from a video interview with Stephan Pastis" by Reading Rockets (2016).
22 Rushkoff (2011), p. 16.
23 McLuhan & Fiore, op. cit., p. 124.

Digital Communication Etiquette

Professional correspondence is a fact of life, whether it's with a teacher, potential employer, supervisor, or colleague. Writing well in an email tells the receiver that you're attentive, professional, and have a stake in the outcome of the conversation.

Here are some basic rules:

1. Grammar, spelling, capitalization always count
   - Repeat until you get it: Email is not text messaging.
   - If your email program doesn't check grammar and spelling, and you know this is a weakness for you, type important emails in Word first.
   - Watch out for autocorrect, especially if you are typing on a tiny virtual keyboard. Your messages can come out comically—or tragically—wrong.

2. Have a concise subject line
   - Blank subject lines signal a lack of importance. They show laziness on the part of the sender and add work for the receiver.
   - What's the subject/topic of this message? Be specific.
   - Change the subject line when the topic changes. This makes it easier to find, save, and delete (aka, manage) emails.

3. Format for the medium
   - Emails are not short novels, they are memos and should be structured as such.
   - *Short* paragraphs, one idea per paragraph, no more than one or two sentences.
     o Think about the device on which people will be viewing your message: If it's on a mobile device, long paragraphs are daunting and scream "don't read this now!"
   - Double-space between paragraphs for readability.

     –   Include a greeting, end with your name.
         o It's jarring to open an email and be thrown right into someone's thought process. A simple "Hello" is considerate and helpful.
     –   Say please and thank you. A lot.

4. One topic per email
     –   If you have several topics to cover, send separate emails for each.
     –   An email with multiple questions to be answered is most likely to result in one question being answered and the others overlooked, ignored, or forgotten.
     –   Again, keep messages short. Be aware that people may be receiving hundreds of emails per day. (You may already be experiencing the onslaught, so you know.)

5. Personal versus professional
     –   Professional emails are polite, short, and to the point.
     –   Email should always be formally informal.
         o Use correct capitalization, spelling, and grammar. Take the time to look it up if you are not sure.
     –   Consider your audience. If you are emailing your boss, a potential employer, or your professor, err on the side of being overly formal and respectful.

6. Signature lines
     –   This is where you tell people who you are and provide contact information. You should do this *once* in a **thread**—a conversation that goes back and forth on one topic.
     –   Keep it reasonably short. No need to include more than your name, title, company affiliation, phone number, and email address.
     –   Skip the clever saying. It doesn't say about you what you think it does.
     –   Note: The signature lines come *after* you sign your name—the name you want people to use when they respond to you.

7. Email is intrusive
     –   Business managers learn not to reflexively take something from people when they try to hand things to you. Once you literally take something into your hands, it figuratively becomes your responsibility.
     –   Email is like that. Open an email and it becomes your responsibility. Respond, don't respond, it's now your choice to make.
     –   Respect how much information comes at *everyone* every day.
         o Everyone receives a great deal of email. If you are asking for something, remember to be conscious of the person's time and be grateful for the attention. If you have to send a reminder, you should still be polite.

8. Before you hit "send"
     –   Be sure you are sending the email to the right person.
     –   Do not hit "reply to all" unless you want an e-mail to go to everyone the original was sent to.
     –   Don't email when you're angry or upset. (How could that go wrong?)
         o Write the email *without putting anyone's name in the "To:" slot*, save it, and review it a few hours later. Odds are you'll feel better for getting the frustration out in writing and will be able to craft a calm, productive response.
     –   Always be aware that *any* digital communication can be forwarded around the globe in a matter of moments.

9. Spam is super annoying
    –   If you are emailing more than a few people, put your name in the "To:" slot and put everyone else's email addresses in "BCC," which stands for "blind carbon copy." It means that no one else can a.) see who else is included in the conversation; and, b.) harvest the list of addresses for spam.

10. Managing the monster
    –   Start the day with email triage. Delete junk, flag stuff for follow up, read and delete whatever you can.
    –   Set up topic folders and move important emails into the appropriate topic folder.
    –   If you *send* an important message, move it from your "sent items" folder into a topic folder.
    –   Keep your professional email account professional.
    –   Create two personal email accounts:
        o   One to use for personal correspondence; and,
        o   One for online orders, registration confirmation, etc. This will keep promotional emails to a minimum in your other accounts.
    –   Sort and delete. Sort and delete. Sort and delete.

11. Texting and Group Chat
    –   See above for some basics
    –   Texting is becoming more common, and needs to be treated a bit differently
    –   There is lees expectation for formality, since the goal is to keep it brief
    –   But that does not mean you can be sloppy
    –   Be very careful how you participate in group chats, or in communication platforms like Slack, Teams, or Google Chat
    –   As with the communication above, be polite, and remember that what you write in digital communication is easily passed around and shared

12. Zoom and Applications like it
    –   Follow the lead of the situation. If it would seem appropriate to have your camera on, strongly consider doing so. It will help you stay engaged.
    –   Turn off self view. Studies show that seeing yourself staring back at you for long periods of time can result in Zoom Dysmorphia. In short, you may fixate on your personal appearance, which can result in anxiety. Not healthy.
    –   Look at your background. Is it appropriate for the situation? Is there anything in the frame that you might not want shared across the internet? Challenging picture in the background? Lighting that makes your face a silhouette?
    –   Update your name so it displays the name you prefer.
    –   Mute your mic if you are not talking, especially if it would add noise that everyone will have to put up with—barking, tweeting (from real birds), construction, blenders, etc.
    –   Keep in mind that even when you chat to a specific other person in the Zoom session that it might be seen by others sooner or later.
    –   Don't multitask. It's really easy to tune out when you're in Zoom with your camera off, because no one can see you doing other things. Which brings us to …
    –   Don't do anything you wouldn't do in public. There are an unfortunately high number of instances where people have forgotten to mute the video and/or sound and gotten themselves into serious trouble.
    –   Be gracious to hosts and other guests. Think of how much difference it has made to you on Zoom when you can tell that people are engaged with what you are saying.
    –   Don't use Zoom video filters or avatars. They're distracting.

Professional Use of Social Media

These tips are the author's condensed version of two weeks of hanging on every word uttered by social media goddess Alexandra Manzano Fransway in the newsroom of *The Oregonian* in Portland, Ore.

Explore

Check out as many social media platforms as possible (10 is a good starting point). Then pick three to focus your efforts on. Become an expert on what those platforms have to offer and how to use them.

Research

Learn as much as you can about your preferred social media platform. This is what search engines are for.

Read

The best way to learn how to use a site well is to follow someone who uses it well. Doesn't hurt to follow someone who doesn't use it well, too. We learn a lot from the mistakes of others.

Set a goal

And stick to it. Three posts a day? One a day? Keep your feeds active.

Be the brand

What you post represents your "brand." If someone searches for you, what does your social media "resume" tell them about you?

Don't be overly random

Social media enables you to have multiple identities that focus on particular topics. Passionate about your profession? Running? Backpacking? Establish yourself as someone that others want to follow and share content from. Occasional randomness is OK, but focus is better.

Follow and tag

Want to build a following? Tag people. Share. Give attribution. Follow people who are considered experts in what you do.

#@&*

Understand how each social media site you use works. Don't "get" hashtags? Don't know what that period means at the start of a Twitter post? Not sure how or why you should tag people or give attribution? That's what search engines are for.

Be flexible

Professional use of social media also requires you to be able to use a smartphone well—searching the news, updating blog entries, live tweeting quotes, describing crowd reactions, taking photos and video. You need to know how to capture a moment for people not there.

Use common sense

Think before you post and review what you've chosen not to post. Have a purpose.

Self-edit

A lack of attention to grammar and spelling signals a lack of credibility.

Get to the point

All social media posts should be short.

Post sparingly

Don't clutter the feed. If users think you are over posting, they are more likely to drop your content from their feeds. Think about what—of all the information you consume in a day—you would want to tell someone at home at the end of it.

Read

Don't skim an article or skip through a video, then post and/or share. Be sure you understand what you are posting.

Oops

If you make a mistake, own it. It's going to happen. Be ready to explain what you can do to prevent future errors.

Voice

Professional, but conversational. Roughly 75% of what you post on social media should be *valuable*; 25% should be *personable*. All of it should be professional.

Balance

Think about your social media strategy and how to keep your personal and professional personas separate. It's very difficult. A lot of journalism now is going to require you to be a more public person.

Protect your privacy

Update your privacy settings. Frequently.

If you post photos to a public feed, post general photos and videos. Save personal stuff for protected space, private accounts.

Protect your mental space.

Don't over-engage in social media on evenings or weekends. Give yourself room to recover from information overload. Embrace the physical world for a refreshing change.

Stop and SIFT.

Regarding social media, when you find something, don't just repost without checking. The SIFT method (covered in Chapter 6) encourages you to:

- **Stop**:
- **Investigate** the source:
- **Find** trusted coverage: and
- **Trace** to the origin.

And you do this with the FOUR MOVES:

1. **Check for previous work** by looking at fact checking websites, for example, to see if the item in mind has been fact-checked. Example sites include Snopes, PolitiFact, Factcheck. org, NPR Fact Check, and Hoax Slayer. You can find these sites pretty easily, but you still have to be responsible for knowing what that site does and how it works.

2. **Going upstream** means seeing where the information came from, going to the original source. This can also require that you understand how news and information works in real time. Many rumors and hoaxes arise out of news coverage of a breaking story that reports a bit of information it later retracts. Sadly, people feel compelled to spread the errors in breaking news coverage … any maybe even start to think that the retractions are part of a conspiracy!

3. **Reading laterally** means to go out of the site where you found the information and look at how others are covering it. It works in tandem with going upstream, since the idea is that multiple places offer more solid confirmation.

4. **Circling back** is a good strategy if the checking process gets you lost, or worse, following down a rabbit hole. You can go back to where you started with the new information you gained from the first run at checking, and find other search terms or other places on the web that can confirm or refute the idea.

Writing for Digital Media and Content Credibility Cues

Online consumption is characterized by shallow reading, combined with selective depth. It's a hunt for information, in which users skim content until they find what they're looking for. And, even when they do pause to evaluate information, they don't read—they continue to skim.

So how do people read on digital media? They don't.

Digital media users skim. And, note, we're not calling them "readers," they're *users*. Digital media summons an entirely different behavior pattern.

Studies have shown that reading comprehension on desktop screens is lower than 40 percent. On mobile it drops to about half of that.[1]

Content for digital consumption must be formatted differently. It must be scannable. Here are some tips for making longer stories, blog posts, articles, etc. scannable, skimmable, and quickly consumable:[2]

- A strong **headline**

- A short **summary**
  o No more than one or two sentences that explain the gist of the content.

---

1 Media guru Jakob Nielsen of the Nielsen Norman Group is a wealth of information on this topic. Go to his website, nngroup.com, and search for "mobile comprehension" and "writing for the Web." You'll find a tremendous resource on digital media use and engagement.
2 Same Web source, nngroup.com. Search for "How users read on the Web." You'll find an article by Jakob Nielsen from 1997 that may be more relevant than ever. We've tried to follow Nielsen's guidelines in this book through shorter paragraphs, breakouts, and a more conversational tone. It's how you read. It's how we've all been trained by the Web to read.

- Bulleted **lists**
  - o Ideally these are simple lists of the subheads and hypertext that links to the anchors in front of each subhead.
  - o In other words, when you click on one topic on the bulleted list, it takes you to that part of the story.

- A **byline**

- A **timestamp**
  - o This tells when the content was posted and allows the user to instantly assess if it is contemporary or not.

- Thematic **chunking**
  - o Break the story into logical sections, separated by subheads.

- Meaningful **subheads**
  - o These should appear every few paragraphs and include words that explain what the following section covers.
  - o Each subhead should have an anchor, which is a little piece of HTML code that allows you to directly hyperlink to the subhead.
  - o This allows a search engine to match the subhead to a search string and send a user directly to the content that interests them without having to wade through the rest.

- **One idea** per paragraph
  - o Users will skip over any additional ideas if they are not caught by the first few words in the paragraph.

- Highlighted **keywords**
  - o These should match up with terms you think someone would put into a search engine to find your blog, article, story, etc.

- Use **inverted pyramid** style
  - o Most important details first. If a user is engaged, he or she will keep reading.

- **Half the word count** of conventional writing
  - o Long, in-depth stories have a place on the web, but the length of the story should conform to the importance and complexity of the issue.
  - o No one is going to tell you to write longer—and if they want to know more, well, that's what your contact information is for.

- Contact **email** and **commenting**, to allow the conversation to continue

Site Credibility Cues

An important part of digital media literacy is being able to quickly assess whether Web resources are credible or not. And knowing site credibility cues helps you create content that appears credible to others.

Whether you're talking about apps or websites, digital is a highly competitive medium. People have millions of other places they can be (as Google's search return count will tell you) and tend to rely on first glance to determine whether to dig farther into a site. Here's a basic outline of the **credibility cues** you should look for before you even begin to consider content:

1. **Design**
   - How does the site *look?* Does it use high-quality graphics? Is it aesthetically appealing? Do all elements work in harmony?
   - Sites that look like a ransom note rarely possess the needed credibility cues that get users to the content.

2. **Information architecture**
   - This signals attention given to the basic structure of the site. If you can't grasp the purpose of a site with a quick **F-pattern** glance (top-to-bottom, top-to-right, mid-left-to-middle), it's likely you won't be there long.
   - Can you find the navigation bar? Does it appear that the site will be easy to use, something that requires a surprising amount of thought.

3. **Use of hypertext**
   - A key part of the World Wide Web is its ability to link pages together by creating a connection between Websites. Hyperlinks require the use of Web addresses, which are technically are called **Uniform Resource Locators**. When you click on a hyperlinked URL, it takes you to the page.
   - But a hyperlink is not an indicator of credibility, *hypertext* is. The difference lies in the ability to integrate links into the content narrative. Words and phrases that are hyperlinked are called hypertext.

4. **Source**
   - Who created this site? Look at the bottom of the screen, which is where copyright and site origin info often hides.

5. **Outbound links**
   - This is known as **associative linking**. The pages a site links to say something very important about the site itself. Hypertext should complement a site's purpose, not contradict it.

6. **Grammar, spelling, typos**
   - It's this simple: Find a major typo, leave the site and don't return. A site that passes the first four credibility checks, but has a screaming typo is indicative of a good designer. While the site may look great, no care is being given to the content.
   - Find a minor typo, consider how well the site adheres to the previous cues. Find a few and you have to give up on the site. Fail.

Putting the Cues to the Test

Let's take a practice run at applying credibility cues to a website. Without even getting to content, we can learn a great deal.

Say you're doing research on Martin Luther King, Jr. in preparation for Martin Luther King Day. You head for The Google and put in the search term, resulting in more than 64 million sites that contain that name. The first is Wikipedia. The second may be martinlutherking.org. Ah. Sounds perfect.

You follow the link for martinlutherking.org and land on a site with a dated design, but the color scheme basically works and there is hypertext. The source info for site ownership appears at the bottom of the page, as is often the case: "Hosted by Stormfront." This hypertext is an outbound link that tells you everything you need to know about this site: Stormfront's motto is "White Pride World Wide."

Yes, Stormfront is a neo-Nazi organization. And, while the site noted above was a racist propaganda source for many years, it's currently offline. But it may exist in another incarnation, so use care.

The nature of digital media use is we don't read; we skim. This site would pass a skim test, which is likely the creator's intent. Mindlessly linking to a white supremacist site—or any other ideology bound content that doesn't reflect your beliefs—is likely to have a direct impact on your value as a resource to others. Carefully reading the credibility cues of a site before citing or linking to it can *save your credibility* (and career).

The Care and Reading of a URL

A site's web address, officially known as a **uniform resource locator**, can tell you things that the rest of a site doesn't. We all know what a URL looks like, but what you may not know is the important credibility information it provides.

Let's take the following link apart, piece-by-relevant-piece:
https://www.csuchico.edu/jour/department/faculty.shtml#wiesinger

- **Protocol:** http or https
  o Hypertext Transfer Protocol, the protocol that is the World Wide Web.
  o An "s" tacked on to the end indicates an encrypted, secure connection—technically a secure sockets layer—between your computer and the website.

- **Domain name:** csuchico
  o This is the root site, the host for the content. Sometimes that is a business or organization, sometimes it's a platform service provider (like WordPress).
  o The combination of the domain name and top-level domain should always take you to a site's homepage.

- **Top-level domain:** .edu
  o Only .edu and .gov have limited use in the United States. Neither can be used by any organization outside of education for the former and federal government for the latter.
  o Other top-level domains used in the U.S. include .com (commercial), .net (community network), .org (organization). These don't necessarily indicate anything, as many corporations and organizations pay for the right to use as many top-level domains as possible.
  o Need an example? Check out the difference between whitehouse.gov and whitehouse.net.

- o **Country codes**
  - ▪ Every country in the world has a top-level domain specific to it. For example, France is .fr, Australia is .au, and Ukraine is .ua.
  - ▪ For many countries, these are the only top-level domains used. In France, Google is google.fr, rather than the google.com you're accustomed to in the U.S.
  - ▪ Anyone can use country codes—they're not restricted. Many companies use them to customize and shorten long URLs. For example, The Oregonian newspaper in Portland has used orne.ws, shorthand for Oregon news (OR news, get it?). The .ws is the country code for Western Samoa.

- **Path directory:** jour
  - o OK, I'm on the csuchico domain, now take me to a department within the university.

- **Subdirectory:** department
  - o This is simply a file folder, or a directory, that exists within the path directory.

- **File:** faculty.shtml
  - o The file is in Hypertext Markup Language, which must be decoded by a browser.
  - o The "s" means it has a Server Side Include, which means that there are things that the host site is including in the page that are not available for editing by the person who posted the content.
    - ▪ SSI content often relates to the overarching organization and can include things like company logos and branding, directories, site search, etc.

- **Anchor:** #wiesinger
  - o Want to link directly to something within a given page? Create an anchor.
  - o Anchors are little pieces of HTML code that are placed before whatever you want to link to. You then have something to actually, well, *link to*.
  - o In this example, there's a list of hypertext faculty names at the top of the page and when you click on "Susan Wiesinger," the link takes you to the anchor, which is beside information about her.
  - o Can also be used to create links to the top of the page. In that case, the anchor would be placed at the top of the page, perhaps called #top, and hyperlinked to words that tell the user where the link is taking them, like "back to top."

File Management

File Management 101: Saving Stuff

Take a look at your computer's desktop. Go ahead. We'll wait.

Is it covered with individual files? Do you save everything there so you can find it?

A desktop full of files and icons doesn't slow down your computer, contrary to popular belief. What it does do, however, is slow down your ability to find anything.

It's like taking all of your clothes and dropping them on the living room floor, along with all your shoes, CDs and albums of your favorite music, DVDs of your favorite movies, and prints of every photo you've ever taken. Mix in the tools every home should have, hammer, screwdriver, scissors, duct tape, etc.

Now try to find something.

Want to find that blue shirt your aunt gave you for your birthday? The running shoes you haven't worn since last month? You have to hunt through the entire pile. Not efficient, not practical, not happening. And, let's be honest, would you even remember what was in the depths of the pile? Probably not, which is how duplications—and mistakes—happen.

This is why we have separate spaces for separate things in the physical world—closets, dressers, shelves, cabinets, and photo albums. A place for everything and everything in its place.

Think about the stuff you use every day in your bathroom. That stuff is not going to end up on the living room floor because you need it where you need it. You put your shampoo and soap in the shower, your toothpaste and toothbrush by the sink, and your hairbrush on the counter in front of the mirror. You keep those things where you use them.

Now let's apply this to file management. Let's pretend your computer is your home. Things that go together should be saved in one place together.

One of the first things a new computer asks you to do is name your digital home, aka the section of the hard drive where you'll store your stuff. Hard drives typically have lots of sections, places for applications, common folders (documents, downloads, music, pictures, etc.) Your root directory typically is your name, maybe first and last, maybe just last.

So let's ignore the common folders for a moment and focus on your root directory. This is the directory where most of your stuff should end up.

Inside that, you would make a subdirectory that represents each room of your figurative house. Within each of those directories you would make sub-subdirectories that represent each place you can store something. Within those sub-subs, you put your relevant files: Shirt files into the closet directory; underwear and sock files into the dresser directory. Got it?

Your desktop is more like your digital living room. This is the space you see every day and others can see when they glance at your computer screen. You want it to be tidy.

The documents folder, a common, default folder where files go to die, is akin to your garage. This is where you save a bunch of stuff that you're just sure you'll organize later. It can be very hard to find anything in this catch-all space. Instead, save the stuff where you can find it. Just like you do in your physical world.

File management is a hierarchical process, one in which you choose what your categories are and then sort things into them. The next time you need to find a file, say for school, you go to the directory called "school" and start looking. Within that directory you might have subdirectories for each class, with sub-subdirectories of individual assignments within the class. Got it?

It comes down to this: Make folders and put stuff in them. Neaten up that desktop already!

File Management 102: Naming Standards

Sticking to some directory and file naming standards will make life much easier, trust us. This is particularly true if you are naming files to add to a website. Plus, consistent file naming helps you find stuff. Here are some very basic naming standards to follow:

- **Keep file and folder names short**
  - o About 8 characters are perfect. This isn't dictated by your computer—the names can be as long as you want them to be. But the best practice is to keep them short.

- **No symbols**
  - o Symbols are things like $%#@*&. Also, apostrophes. They also are codes and mean something to your computer. Don't use them.
  - o Numbers are OK.

- **No spaces**
  - o Spaces are characters, characters are code. Spaces will show up in a web address as either + or %20.
  - o Some non-space options:
    - *camelCase*
      Separating multi-word names with capital letters, rather than a space. (kauiBeachSunset.jpg, like the humps on a camel)
    - *kebab-case*
      Separating multi-word names with hyphens. (kaui-beach-sunset.jpg, like a shish kabab stick running through the words)
    - *snake_case*
      Separating multi-word names with underscores. (kaui_beach_sunset.jpg, like a snake slithering along the ground below your words)

- **Leave extensions**
  - o They tell your computer what to do with a file.
  - o For example: .jpg is an image, .docx is a Word file, .html is a web page, etc. Every file must have an extension to be read by computer programs and the Web.

Key concepts

Index

Made in United States
North Haven, CT
14 February 2024

48720081R00137